KEY LEAVING C
BIOLOGY DEFINITIO

Coming to Terms with
BIOLOGY

Declan Finlayson

First published 2005

The Educational Company of Ireland
Ballymount Road
Walkinstown
Dublin 12

A trading unit of Smurfit Ireland Ltd

Approved ✓ Quality
System

The
paper used in
this book comes
from Managed
Forests in
Northern
Europe
For
every
tree
felled, at
least one new
tree is planted

Illustrations: Michael Phillips
Design and layout: Design Image
Cover Design: Design Image
Cover photograph: Gregor Mendel. Image courtesy of Corbis

Printed in the Republic of Ireland by Future Print.

0 1 2 3 4 5 6 7 8 9

Foreword

This book and CD is an essential companion for all students of Leaving Certificate Biology. Clear and concise definitions ensure this package is invaluable, both as a study aid and for exam preparation. The new Biology syllabus introduced new terms. These have been included, as well as references to the more recent developments in modern Biology.

The Book
- Contains over 1,200 terms.
- Explanations are clear and concise.
- Illustrated diagrams throughout.
- User-friendly cross-reference system is highlighted in **bold type**. The highlighted words are defined elsewhere in the book.

The CD
- Specific definitions mentioned in the syllabus and related terms are highlighted in red.
- Contains all terms in the book, plus 1,400 more.
- Illustrated colour diagrams throughout.
- All terms on the CD are hyperlinked. This allows for rapid cross-referencing.
- Minimum computer skills required for use.

Every effort has been made to ensure the accuracy of the text and hyperlinks. Any inaccuracies in the book or CD can be reported to the author at biologydictionary@eircom.net

I would like to thank the following people, without whom this book and CD would not be possible – my Leaving Certificate Biology students, my colleague John O'Sullivan, my sons Ian and Ken, and especially my wife, Nora, for her monumental patience from conception to completion of this book and CD.

I wish you every success in your Leaving Certificate Biology examination. Enjoy, have fun and learn.

Declan Finlayson

Aa

abdomen	1. Lower section of the trunk of a body, below the diaphragm. See the respiratory system for diagram. 2. End section of insect.
abiotic environment	Non-living physical elements of the environment e.g. stone, water, light intensity, pH etc.
abiotic factors	These are the non-living features of an ecosystem that affect the community. They consist of the physical and chemical conditions, and they vary between ecosystems that are terrestrial or aquatic. They include: temperature, light intensity, air speed, water current, humidity, pH, dissolved oxygen, salinity, nitrate, phosphate and other plant nutrients.
abscisic acid	Growth inhibitor found in plants. It is a plant growth regulator (hormone) that promotes dormancy and leaf fall (abscission), and inhibits longitudinal growth.
absorption	The taking in of heat, fluids, gases or nutrients by a cell through its membrane. In digestion, it is the taking in, by the blood and lymph systems (lacteals), of the end products of digestion (monosaccharides, amino acids, fatty acids, glycerol, etc).
accommodation	Adjustment of the curvature of the lens of the eye to focus on objects at different distances i.e. it can focus on near or far objects (not at the same time). The ciliary body and suspensory ligaments control the curvature of the lens.
acetylcholine	A neurotransmitter i.e. a chemical compound (substance) that transmits a nerve impulse from one neuron (nerve) to another. It is found at the synapse and is discharged after transmission of the impulse.

acetyl co-enzyme A	An intermediate product formed during the aerobic respiration of a molecule of glucose. Pyruvic acid loses one molecule of carbon dioxide and two hydrogen atoms to form acetyl co-enzyme A. The acetyl co-enzyme A then enters Kreb's cycle and reacts with citric acid to form oxaloacetic acid.
acid	A substance that reacts with bases to form salts and water. All acids contain hydrogen, which can be replaced by positive metallic elements or radicals. Acids are also called proton donors i.e. they give away a hydrogen ion (H^+), which is a proton. Acids increase the number of hydrogen ions in a solution. Can be sharp or sour tasting. See also pH.
acid rain	Acid rain refers to very acidic rain with a pH of 4.5 or less. Sulphur dioxide (SO_2) dissolves in rainwater to form sulphurous acid (H_2SO_3) or reacts with particles in the air to form sulphuric acid (H_2SO_4).
acquired immunity	Immunity induced by inoculation (injection). Or Once infected by micro-organisms, antibodies are formed which kill the micro-organisms and prevents further infection.
acrosome	Found at the tip of the sperm above the nucleus. It contains enzymes that break down the jelly coat surrounding the female egg thus allowing the sperm nucleus entry into the egg. See sperm for diagram.
activation energy	The minimum amount of energy needed to start a chemical reaction. This varies from reaction to reaction.
active	Usually proceeds a process e.g. active transport. Active processes require energy, in the form of ATP, released from glucose in the mitochondrion during the process of respiration. Oxygen is required.

active immunity	Long-term immunity resulting from (a) inoculation (the injection) of vaccine containing a weakened strain of a pathogen or its toxin, (b) an infection by a pathogen. The body produces an antibody in response to the presence of either of these and retains the ability to produce it again if attacked by the pathogen at some later stage i.e. it is long-term immunity. Compare passive immunity.
active site (of enzyme)	The position on an enzyme where the substrate is attached. See enzyme-substrate complex for diagram.
active transport	The movement of substances (solutes or ions) from a region of low concentration (hypotonic) to a region of higher concentration (hypertonic) against the concentration gradient through a semi-permeable membrane (cell membrane). It requires the use of energy (in the form of ATP) by the cell. ATP is a product of respiration. Respiration usually requires oxygen and occurs in the mitochondria. An example of active transport is reabsorption in the nephron of the kidney.
acute	Referring to pain or disease, it means intense, coming quickly to a crisis. Compare chronic.
adaptation	Process or change or method by which an organism or species becomes adjusted to its environment. This is one of the main points of Darwin's theory of evolution by natural selection: organisms adapt to their environment and organisms best adapted will have a greater chance of surviving and passing their genes on to the next generation.
adaptive radiation	The different functions of the same structure shows adaptation of a common structure e.g. the forelimb.
adenine	Compound containing nitrogen. It is one of the purine nucleotide bases of DNA and RNA, also part of ATP molecule: adenine + ribose = adenosine.

adenovirus	A DNA virus that causes infections of the tonsils, intestinal tract, respiratory system and the eyes (conjunctivitis).
ADH	Antidiuretic hormone. This hormone, produced by the posterior lobe of the pituitary and secretion of which depends on the water content of the blood, encourages the reabsorption of water in the nephron of the kidney. Controls osmoregulation. A deficiency causes diabetes insipidus. ADH is also known as vasopressin. Note: alcohol inhibits ADH production i.e. if there is no ADH vast quantities of watery urine will be produced ⇒ more urine coming out than beer going in.
adhesion	The ability of the molecules of one substance to stick to a different substance.
adipose tissue	Layer of fat storage cells beneath the dermis. Its functions are energy storage and thermal insulation. See skin for diagram.
ADP	Adenosine diphosphate. It is formed by the release of energy in a cell from ATP. It is converted back into ATP by the addition of a phosphate group during respiration.
adrenal gland	Ductless glands (endocrine), one found on top of both kidneys. The adrenal glands secrete adrenaline.

adrenaline	Hormone produced by adrenal glands, the 'fight or flight' hormone. It stimulates the body in emergencies. Promotes respiration (with the release of energy) and circulation.
adventitious root	Root that does not develop from the radicle e.g. (a) climbing roots of ivy, (b) roots of cuttings that arise from a node.
aerobe	See aerobic bacteria.
aerobic bacteria	Bacteria that can only live with or in the presence of free oxygen from the air. They require oxygen for respiration.
aerobic respiration	The controlled release of energy from food within a cell using oxygen. The process is controlled by enzymes and is very efficient. See respiration. $$C_6H_{12}O_6 + 6O_2 \rightarrow ENERGY + 6H_2O + 6CO_2$$
afferent arteriole	Blood vessel (small artery) bringing blood towards the glomerulus of the nephron in the kidney. See Bowman's capsule for diagram.
afferent neuron	A nerve conducting (carrying) an impulse (message) inwards or towards the central nervous system (CNS = brain and spinal cord). Cell body found somewhere along length of neuron, outside CNS e.g. sensory neurons are afferent neurons. See neuron for diagram.
afterbirth	Placenta, membranes that surrounded the foetus and remains of umbilical cord discharged from the uterus after the birth of the baby.
agglutination	Clotting of blood to seal a wound.

AIDS	Acquired Immune Deficiency Syndrome. This is a collection of disorders that develop as a result of being infected with HIV. *Acquired* because you have to 'catch' it, i.e. you don't inherit it; *Immune Deficiency* because it lessens your defence against disease; *Syndrome* because of the wide variety of ailments associated with it and not just a specific illness.
air sac	See alveolus.
albumen	Simple protein found in blood, milk, the white of an egg and muscle.
alcohol	Ethyl alcohol or ethanol (C_2H_5OH). It is one of the end products of anaerobic respiration (fermentation), the other being carbon dioxide.
alga(e)	Plants that do not produce flowers or seed, mainly aquatic, sometimes filamentous. No true roots, stems or leaves e.g. seaweeds.
algal bloom	Scum formed by vast numbers of algae on the surface of lakes and slow moving streams, usually in spring and autumn. Caused by high mineral content in the water.
alginate	A component of the cell walls of many types of seaweed. Alginates have an affinity for water, and so prevent desiccation of the seaweed when the tide is out and they are exposed to the air.
alimentary canal	Tube that transmits food through the body of an animal from mouth to anus. See digestive system for diagram.
alkali	Same as a base i.e. reacts with acid and forms a salt, but also soluble in water to give hydroxide ions (OH^-) e.g. NaOH, KOH, etc. See also pH.
alkaline	Describing an alkali.
allele(s)	Alternative forms of a gene or a pair of genes found at the same locus/position on homologous chromosomes controlling the same trait.

alveolus	(Plural = alveoli.) It is an air sac, a small terminal cavity found at the end of the bronchioles of the lung where gaseous exchange (carbon dioxide and oxygen) between air and blood takes place. The gaseous exchange is rapid because: • there is a huge number of alveoli, about 300 million in each lung = area of 70m^2 • each alveolus wall is very thin, only one cell thick and surrounded by fluid • a huge capillary network surrounds each alveolus • the capillaries are only one cell thick.

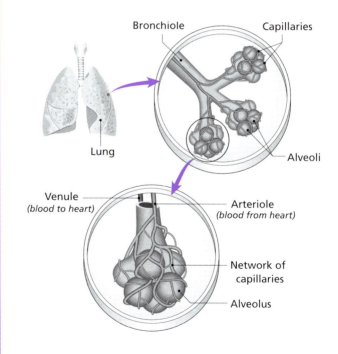

Bronchiole Capillaries

Lung

Alveoli

Venule
(blood to heart)

Arteriole
(blood from heart)

Network of
capillaries

Alveolus

amino acid	Structural unit of proteins, containing an amino group ($-NH_2$). There are twenty common amino acids found in proteins. More are found in cells and tissues but not in proteins.

ammonia	NH_3. A pungent smelling, colourless gas, soluble in water forming ammonium hydroxide (NH_4OH). Formed in liver as a result of deamination of amino acids. It is very poisonous, and is converted to the less toxic substance urea.
ammonification	The conversion of nitrogen compounds (e.g. urea and protein as a result of excretion and death respectively) into ammonia. See nitrogen cycle.
amniocentesis	A method of testing the foetus in the uterus (womb) for chromosomal disorders. A sample of amniotic fluid is withdrawn from the uterus using a needle and the fluid and foetal cells it contains are analysed.
amnion	Fluid-filled membrane surrounding foetus (developing baby) in the uterus (womb). See embryo for diagram.
Amoeba	A single-celled aquatic protozoan with no fixed shape.

Cytoplasm
Ectoplasm Endoplasm
Cell membrane
Pseudopodia
Nucleus
Fat droplet
Food vacuole
Waste products
Contractile vacuole

amylase	Enzyme produced by the salivary glands (salivary amylase or ptyalin), the pancreas (pancreatic amylase) and the small intestine (intestinal amylase). It converts starch to maltose during digestion. Its optimum pH is slightly basic.
anabolism	A chemical reaction which joins small molecules to make larger, more complex ones using enzymes e.g. photosynthesis. Anabolic reactions require energy. See metabolism and catabolism.

anaemia	Condition of the blood resulting from a deficiency of haemoglobin. Symptoms include tiredness, paleness and breathlessness after activity. A deficiency in haemoglobin results in less oxygen being brought to the body cells → less oxygen available for respiration → less respiration → less energy produced → tiredness.
anaerobe (anaerobic bacteria)	Bacteria (micro-organisms) that can or must live in the absence of free oxygen from the air.
anaerobic respiration	The enzyme-controlled release of energy from food in a cell in the absence of oxygen. It is an inefficient process. Oxygen may be present but will not be used. e.g.

(i) in yeast cells – fermentation

$$C_6H_{12}O_6 \rightarrow ENERGY + 2C_2H_5OH + 2CO_2$$

ethyl alcohol + carbon dioxide

or ethanol + carbon dioxide

(ii) in human muscle (results in cramp)

$$C_6H_{12}O_6 \rightarrow ENERGY + 2CH_3CH(OH)COOH$$

lactic acid

anaphase	One of the stages of cell division (mitosis and meiosis) during which the centromeres split and half of each chromosome pair moves from the equator to the poles of the cell. This is followed by telophase.
anatomy	Science of the structure of plants and animals, both internal and external. Dissections of organisms.
angiosperm	A flowering plant. It has vascular tissue i.e. xylem and phloem, double fertilisation, and the fruit produced contains seeds.
annual	A plant that completes its life cycle in one growing season.

anorexia nervosa	A serious psychological condition, usually occurring in adolescent females, in which the patient makes every possible effort to avoid weight gain. Sufferers may starve themselves, induce vomiting, take laxatives and even diuretic tablets. These people have low self-esteem, a false impression of their body size and usually an intense phobia about becoming obese. Severe weight loss often results and death may occur sometimes.
antagonistic muscle pairs	These are **muscles** working in pairs in opposite directions controlling the movement of a **joint** e.g. movement about the elbow controlled by **biceps** (= **flexor muscle** = bring up bones) and **triceps** (= **extensor muscle** = bring back bone).

Tendon — Biceps relaxed — Biceps contracted — Scapula — Triceps contracted — Triceps relaxed — Ulna — Radius — Humerus — Tendon

anther	The part of the **stamen** that produces and stores **pollen** until it is ripe. See **pollen mother cell** for diagram.
antibiotic	Substance naturally produced by living **micro-organisms** (e.g. *Penicillium notatum*) which destroys or inhibits the growth of other **micro-organisms**, especially **bacteria** or **fungi** e.g. penicillin, streptomycin, tetracycline, etc.

The ability to make this substance is controlled by a gene, which can be passed on to other bacteria of the same or different species. Note: antibiotics have no effect on viruses.

antibiotic resistance	If a patient is given an antibiotic it kills all bacteria (good and bad) except those that are naturally resistant to it – they may not be pathogens. These antibiotic resistant bacteria now flourish and increase in numbers. If a pathogen now attacks the patient it can pick up the resistance to the antibiotic from the non-pathogen, thus conferring resistance on itself. The number of antibiotic resistant bacteria is on the increase, so the use of antibiotics should be minimised. See antibiotic.
antibody	Protein substance produced by lymphocytes in the immune system to attack and destroy or counteract antigens (micro-organisms or their products). Antibodies are specific.
antigen	A substance (e.g. a micro-organism or its waste products – proteins or polysaccharides) that the immune system of the body recognises as 'foreign' and as a result will produce antibodies that will react with the antigen and neutralise or destroy it.
antigen antibody reaction	A reaction between an antigen and an antibody resulting in the neutralisation or elimination of the antigen.
antiseptic	An agent that inhibits growth of micro-organisms but does not kill those already growing.
antiserum	Serum from the blood of humans or animals that contains antibodies specific to certain pathogens. When injected into a patient, recovery is almost guaranteed but the patient will not retain any immunity to the disease. The immunity is passive and short-term.

aorta	The main blood vessel leaving the heart and carrying oxygenated blood to the tissues of the body. See the heart for diagram.
apical bud	Bud found at the apex or tip of a shoot.
apical dominance	The suppression of the growth of the lateral buds by the apical bud, thus allowing the stem to elongate.
apical meristem	Meristem (embryonic tissue) found at the tip of the shoot or root responsible for increasing the length of the shoot or root. See root cap for diagram.
apophysis	The swollen tip of the sporangiophore of the black bread mould *Rhizopus*. See *Rhizopus* for diagram.
appendicular skeleton	All bones in the skeleton except the axial skeleton (skull, vertebral column, sternum and ribs). These include wings, legs, and arms or fins and the pelvic girdle and pectoral girdle that join the appendages to the rest of the skeleton. See the human skeleton for diagram.
appendix	A small, dead-end tube arising from the caecum in humans. See the digestive system for diagram.
aquaculture	The growing of plants and the breeding of animals in water.
aquatic factors	Factors that influence the life of the organisms that live in ponds, rivers and seas. For example: depth of light penetration, currents and wave action.
aqueous humour	A transparent liquid found in the eye between the lens and the cornea. Light is refracted (bent slightly) as it passes through this on its way to the retina. The aqueous humour helps to maintain eyeball shape. See the eye for diagram.

arm

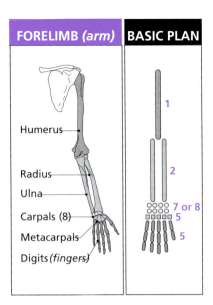

FORELIMB *(arm)*	BASIC PLAN

Humerus

Radius

Ulna

Carpals (8)

Metacarpals

Digits *(fingers)*

1

2

7 or 8
5

5

arteriole | A small artery (less than 0.3mm in diameter).

artery | Thick-walled blood vessel that carries blood away from heart rapidly at high pressure. Blood flows in pulses. Arteries have no valves. They have a small lumen and a three-layered wall:

- outer layer: non-elastic fibres– collagen
- middle layer: elastic fibres and muscles (thick layer)
- inner layer: endothelium – one cell thick.

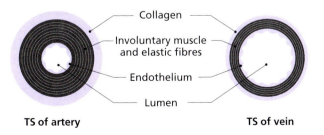

Collagen

Involuntary muscle
and elastic fibres

Endothelium

Lumen

TS of artery   **TS of vein**

arthritis	Inflammation of a joint. There are many pathological (disease-related) causes, including bacterial or viral infection, inflammatory or degenerating disease, commonly rheumatoid arthritis and osteoarthritis. Arthritis can affect different joints, and sufferers may have symptoms of pain, swelling over the joint and restricted movement. The treatment of arthritis depends on its cause – if inflammatory, specific drugs help to relieve pain and swelling. If caused by infection, anti-bacterial drugs are used. Severe arthritis may require joint replacement.
ascending tubule	Forming that part of the nephron just after the loop of Henle through which the glomerular filtrate flows upwards. It is permeable to salt only. See nephron for diagram.
asepsis (or aseptic techniques)	Methods used to prevent unwanted micro-organisms entering an area.
aseptic	Free from all pathogens i.e. organisms that cause disease.
asexual reproduction	This does not involve the manufacture or union of sex cells or gametes e.g. binary fission, fragmentation, spore formation and budding. It involves only one parent and offspring are genetically identical (have the same genetic content) to parent. See vegetative propagation.
aspect	The direction in which something (habitat) is facing i.e. north, south, east, west etc.
assimilation	The process of converting simple nutrients (e.g. amino acids) into cellular/body components (e.g. proteins) i.e. converting non-living matter into protoplasm – an anabolic process (anabolism).

association neuron	Nerve cell that connects a sensory neuron and a motor neuron. See reflex arc for diagram.
asthma	A common breathing disorder caused by narrowing of the bronchioles of the lungs, with accompanying mucus secretion. This can be brought on by the presence of pollen, house mites, dog and cat dander or vigorous exercise. Can be treated with steroid tablets or inhalers and fine nebulised mist (cloudy) solutions for inhalation.
ATP	Adenosine triphosphate. A compound composed of a molecule of adenine, one of ribose and three phosphate groups. The bonds between the phosphate groups are high-energy bonds. When these bonds are broken, energy is released. ATP is the source of energy in a cell – it traps and transfers energy for cell activities. Formed in the mitochondria of a cell during respiration or photosynthesis. ATP is converted to ADP with the release of energy and a phosphate group: $$ATP + H_2O \rightarrow ADP + P, \text{ energy out}$$ The reverse also occurs during its formation: $$ADP + P \rightarrow ATP + H_2O, \text{ energy in}$$
atrioventricular node (A-V node.)	A rounded mass of muscle tissue situated in the septum between the right atrium and right ventricle. It picks up the wave of impulses that cause the atria to contract and transmits them to the ventricles, causing them to contract in turn.
atrium	(Plural = atria.) One of the two upper chambers of the heart that receive blood from the veins. See the heart for diagram.
auditory nerve	The nerve transmitting impulses (created by sound) from the cochlea to the brain. See the ear for diagram.

autoimmune disease	A collection of diseases where the antibodies of the body's immune system attack the host tissue. It is not known why this occurs but drugs, genetic factors or metabolic deficiencies are induced by some conditions. The best-known example is rheumatoid arthritis, when the body's immune system attacks the linings of joints and progressively goes on to attack ligaments and bone.
autosomes	Non-sex chromosomes. 44 of the 46 human chromosomes are autosomes.
autotroph	An organism that manufactures its own food from inorganic sources e.g. a green plant.
autotrophic bacteria	Bacteria that are self-nourishing i.e. capable of making (synthesising) their own food from inorganic compounds.
auxin	A chemical that is a plant growth regulator or a plant growth hormone (e.g. IAA), produced by plants and stimulates cell elongation and cell division in plants. Auxin is involved in tropic responses. It is produced in one place but its action is in another.
A-V node	See atrioventricular node.
available (soil) water	Free water i.e. water in the soil that can be absorbed into roots of plants by osmosis. Due to the adhesive forces of water molecules, some soil water is non-available i.e. the root hairs cannot exert enough osmotic pressure to pull the water off the soil particles.
axial skeleton	This is composed of the skull, vertebral column, sternum and ribs. See the human skeleton for diagram.
axillary bud	A bud found in the axil of the leaf i.e. in the angle between the petiole of a leaf and the stem. See the leaf for diagram.
axon	Long appendage coming from cell body of nerve cell (neuron). Conducts impulses away from cell body. See neuron for diagram.

Bb

B-cell	A lymphocyte responsible for producing antibodies and inactivating antigens by surface recognition. Each B-cell produces just one type of antibody. Produced in the bone marrow, it migrates to and matures in the lymph nodes. On contact with an antigen, B-cells reproduce rapidly. They provide large quantities of antibodies and destroy the antigen. Some of the cells produced remain in lymph nodes as memory T-cells. These provide a rapid response for successive exposures to the antigen – active immunity.
B-lymphocytes	See B-cells.
bacillus	(Plural = bacilli.) Rod shaped bacterium. Bacteria are classified according to their shape.
bacteriophage	A virus that attacks and kills bacterial cells. It is composed of DNA and protein only.
bacterium	(Plural = bacteria.) Microscopic unicellular organisms, prokaryotes in nature. They may be parasites or saprophytes. Some are autotrophic bacteria. Many cause disease (i.e. are pathogenic) or decay and some are used in industry. They are ubiquitous (i.e. found everywhere – salt water, fresh water, terrestrial, airborne).

contd. ...

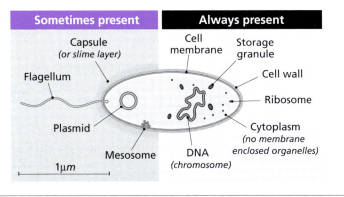

Bacteria are composed of a cell wall, cell membrane, genetic/nuclear material – DNA (not in a nucleus), plasmids, cytoplasm, flagella and the capsule. There are three main types of shapes: cocci, rods and spirals.

Bacteria reproduce asexually by binary fission. Because they have such a short life cycle (about 20 minutes in ideal conditions), numerous generations can be produced in a very brief time. As a result mutations are a very important feature since they can become established in a population very quickly.

balanced diet	One that contains the correct proportions of each of the following: carbohydrate, fat, protein, minerals, vitamins, water and roughage. A balanced diet for an individual will depend on his/her sex, age and level of physical activity. See food pyramid.
ball and socket joint	Type of synovial joint with a rounded end in a concave cup providing great freedom of movement e.g. shoulder and hip joints. See joint(s) for diagram.
bark	Tissues lying outside the cambium in woody plants, consists of dead phloem, dead cortex and cork.
base	A substance which will react with acids to form salts. It lowers the hydrogen ion concentration (pH) in a solution. May have a soapy feel in solution.
base pairs	Refers to nucleotide bonding. In DNA adenine (A) always bonds with thymine (T) and cytosine (C) always bonds with guanine (G). In RNA thymine is replaced by uracil.
batch processing	The organisms being grown are in a five-phase growth curve. Compare continuous flow processing. See growth curve for diagram.
behaviour	The response of an organism to changes in both its internal and external environments.
benign	Not malignant.

biceps	Muscle on the 'inside' of humerus that raises the lower arm and bends the elbow. See antagonistic muscle pairs for diagram.
bicuspid valve	Also known as mitral valve. This valve is in the heart separating the left atrium from the left ventricle. Has two leaflets or flaps. Prevents blood flowing from the ventricle to the atrium when the ventricle contracts. See the heart for diagram.
biennial	A plant that takes two years (i.e. two growing seasons) to complete its life cycle. In the first year, the seeds germinate and grow into a plant which produces and stores food in an underground perennating organ (see perennation), and the foliage above ground usually dies back in autumn. In the second year, new shoots arise from the underground organ to produce flowers and seeds.
bile	A fluid secreted by the liver and stored in the gall bladder. Contains bile pigments, cholesterol and organic salts. Bile aids digestion by changing the pH of the stomach contents from acid to alkaline, emulsifying fats to fat droplets, and activating pancreatic lipase.
bile duct	Transports bile from the gall bladder to the duodenum.

Oesophagus

Cardiac sphincter

Liver

Stomach

Gall bladder

Pancreas

Bile duct

Pyloric sphincter

Pancreatic duct

Duodenum

binary fission	**Mitosis. Nucleus** (or **DNA** in **bacteria**) duplicates and then the contents of **cell** divide. Cell elongates and replicated **DNA** and contents are pushed apart to opposite ends. **Cell membrane** constricts or **cell wall** forms, resulting in two daughter **cells** identical to, but only half the size of, the original parent **cell**. This type of **reproduction** is found in the **protista** and **bacteria**.

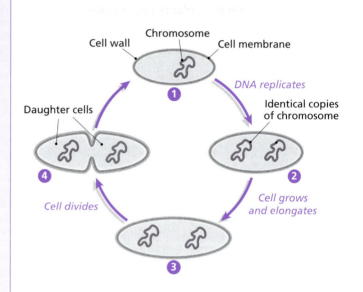

binomial system	System used in naming **organisms**. Each **organism** is given a **genus** name and a **species** name e.g. *Homo sapiens* = human.
Biochemical	See **biomolecule**.
biodegradation	The process of decomposition by living **organisms**, usually by **bacteria**.
biodiversity	Variety of living **organisms**. This can be measured in terms of genetic, **species** or **ecosystem diversity**.
biological key	Series of groups of questions used to identify one feature at a time of an unknown **organism** which eventually leads to its identification.

biological magnification	Pollutants tend to concentrate as they are passed through a food chain. The organism at the end of the chain suffers the highest concentrations. This is biological magnification.
biology	The study of living things. The science that deals with anatomy (dissection), morphology (form/shape), physiology (functions), behaviour, distribution and origin of plants and animals.
biomass	The total quantity or weight of a species per unit area or volume or at any level in a food chain. This term is used when expressing population density and may be measured as either living mass or dry mass (dry weight).
Biomolecule(s)	Substances (molecules) made inside organisms, also known as biochemicals. There are four types of biomolecules found in food: carbohydrates, fats and oils (lipids), proteins and vitamins.
bioreactor	A container in which a living thing is used in the production of something useful e.g. during fermentation using yeast, yoghurt production using bacteria, etc.

Tap — Motor
Substrate in →
Foam breaker —
Substrate and micro-organisms —
Stirrers —
Sparger (to form air bubbles) —
Gas out →
Product out →
Air in →

biosphere	That part of the earth inhabited by living organisms, including land, ocean and the atmosphere in which life can exist. It is the global ecosystem.
biotechnology	Use of living organisms, their cells or products in food production or other substances which are useful to humans e.g. yeast in making alcohol, bacteria in yoghurt manufacture, etc.
biotic	Refers or relates to life or living things.
biotic environment	The living environment, consisting of plants, animals, bacteria and fungi.
biotic factors	These are the living features of an ecosystem that affect the other members of the community. These include: plants for food and shelter, predators, prey, parasites and pathogens, decomposers, competitors and pollinators.
birth canal	The vagina. See female reproductive system for diagram.
birth control	Limiting the number of children born. See contraception.
bladder	A membranous sac filled with fluid or air. • In animals it serves as a reservoir for fluids e.g. urine (urinary bladder) or bile (gall bladder). • In algae it holds air and aids buoyancy or floatation e.g. seaweeds. See urinary system for diagram.
blastocyst	A hollow, fluid-filled sphere of cells that develops from the morula. Attached to the inner surface of the outer layer of cells (trophoblast) is an inner mass of cells from which the embryo develops.

Zygote

Mitosis

Day 1

Mitosis

Day 2

Mitosis

Blastocyst

Morula

Trophoblast
Fluid cavity
Inner cell mass

Day 5

Day 3

blind spot	Where the optic nerve leaves the retina of the eye, there are no rods or cones. This spot is insensitive to light, therefore no vision at this spot. See the eye for diagram.
blood	A fluid tissue, a transport system of nutrients in the human body, circulating in arteries and veins. Blood consists of a liquid plasma containing white blood cells, red blood corpuscles (erythrocytes) and platelets.
blood clotting	Coagulation or the conversion of a liquid blood into a solid clot of fibres in which blood cells are caught.
blood group(s)	Any one of the four types of blood into which human blood can be divided according to its compatibility in transfusions. This compatibility is based on the presence or absence of antigens on the red blood corpuscles and antigens in the serum. The main blood groups are A, B, AB and O. When trying to ascertain compatibility in blood transfusions consider only the effect of the antibodies of the recipient on the antigens of the donor.
blood pressure	The pressure of blood in the arteries recorded at the bend of the upper arm. Two figures are recorded: the contraction (systole) of the ventricle and relaxation (diastole) of the ventricle. Normal readings vary with age but systolic reading should not be higher than 140 mmHg, and the diastolic reading should not be higher than 90 mmHg. Readings above these levels constitute high blood pressure (hypertension); very low readings are called hypotension.
blood vessel	Tube through which blood flows. See artery, vein, capillary, arteriole and venule. See capillary for diagram.

bone

A hard rigid connective **tissue** made from **bone cells** (**osteocytes**) embedded in a matrix of **organic** and **mineral matter** (calcium and phosphorus) making up the **skeleton**.

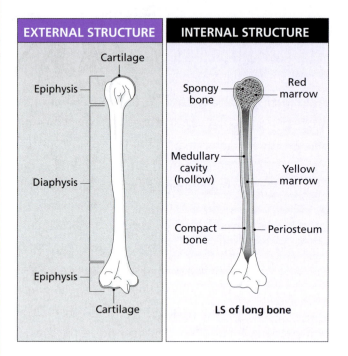

bone marrow

Soft fatty **tissue** filling the spaces in **spongy bone**. Produces **blood cells**. See **bone** for diagram.

| Bowman's capsule | A double-walled, cup-shaped chamber at the expanded end of a nephron (uriniferous tubule), which encloses a glomerulus. |

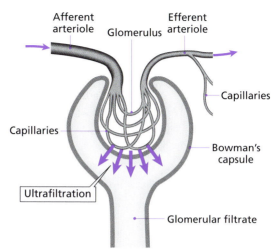

| brain | An enlargement at the top of the spinal cord where most decisions are made in response to information received from the sense organs. In vertebrates, the brain is that part of the central nervous system enclosed in the skull. Consists of the olfactory lobes, cerebrum, pineal body, thalamus, hypothalamus, midbrain, pons, cerebellum and medulla oblongata. |

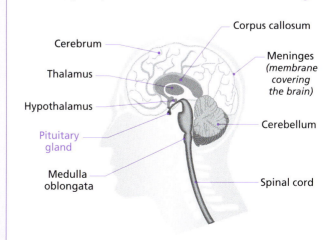

breastfeeding	Suckling a child at mother's breast. Breastfeeding gives a baby the best possible start in life with many benefits: • gives a baby all the **nutrients** it needs in exactly the right proportions for optimum development • is a source of **antibodies** which are passed on to the baby to protect it against allergies and illness • **milk** is always at the correct temperature • encourages bonding between mother and baby, and develops in the mother a high degree of **sensitivity** to her baby. Produces feelings of calm and helps a mother relax • aids recovery of the body after **pregnancy** and hastens the return of the womb (**uterus**) to its original shape and position • can cause a cessation of **menstruation**.
breast milk	Milk produced by the mammary **glands** (breasts) containing the correct balance of **nutrients** and **antibodies**. Best possible food for baby.
breathing	The physical process of taking in **oxygen** (inhalation or **inspiration**) and giving out of carbon dioxide and **water** vapour to the air (exhalation or **expiration**). This is called external **respiration**.

Trachea

INHALATION (active)
muscles contract

Ribs move up and out

Rib

Lung deflated

Lung inflated

EXHALATION (passive)
muscles relax

Diaphragm

Spine

Diaphragm curved down

breathing rate	The number of breaths per minute. The average in an adult human is 16 per minute.
breeding	Propagating, producing offspring similar to parents by selection.
bronchiole	One of the numerous minute divisions of the bronchus. See alveolus for diagram.
bronchitis	An inflammatory condition affecting the bronchi. Caused by respiratory bacteria, viruses and long-term exposure to irritants such as cigarette smoke and air pollutants. Results in a build up of mucus in the bronchi.
bronchodilator	Substance in an inhaler, used by a person suffering from asthma. It widens the bronchioles and makes it easier to breathe.
bronchus	(Plural = bronchi.) One of two main divisions of the trachea (windpipe) leading to a lung. See respiratory system for diagram.
bud	Beginning of a shoot, leaf or flower. Leaf or flower not fully open. See leaf for diagram. Also an asexual growth later separating to form new organism. See budding.
budding	A method of asexual reproduction where a new individual arises from an outgrowth or bud that may or may not separate to form the parent. For example, in

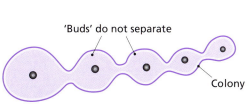

Parent cell   Bud   Daughter cell

Mitosis

'Buds' do not separate

Colony

contd. ...

yeast a bud is formed on the parent cell. It enlarges and fills with cytoplasm. Nucleus divides in two by mitosis and one moves into the bud: this bud may remain attached to the parent and undergo further budding which will result in the formation of a colony or it may detach from the parent, undergo further budding and form a new colony itself.

Or

A method of artificial plant propagation where a bud is attached to the stock e.g. roses.

bulb

Modified bud. Swollen, underground, overlapping fleshy leaves and reduced stem e.g. onion, tulip.

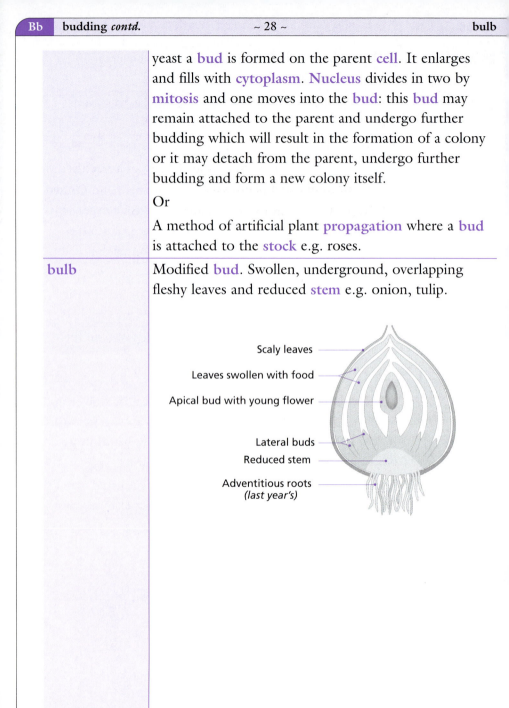

Scaly leaves

Leaves swollen with food

Apical bud with young flower

Lateral buds

Reduced stem

Adventitious roots
(last year's)

Cc

caecum	The first part of the large intestine of mammals. See the digestive system for diagram.
callus	A mass of new tissue produced e.g. during micro-propagation in a test tube or at the site of a fractured bone or injury to a plant, etc.
cambium	A layer of tissue in the stems and roots of dicotyledonous plants from which the annual growth of wood and bark occurs, or gives rise to secondary tissues (xylem and phloem).
cancer	A group of disorders caused by the abnormal and uncontrolled division of cells, resulting from the loss of control over their rate of growth and cell division, forming a tumour which then invades the surrounding tissues.
capillarity	The ability of water to climb in tubes of narrow bore (e.g. xylem) or between tightly packed particles due to its adhesive and cohesive properties, in the spaces between soil particles.
capillary	Connecting link between arterioles and venules – walls one cell thick (endothelium). Substances diffuse from

contd. ...

Capillaries

Arteriole

Venule

Artery

Vein

from heart   *to heart*

the capillaries into the fluid in the spaces between the body cells (tissue fluid or extracellular fluid) and from there diffuse into the cells. Waste products in the cell diffuse in the reverse direction e.g. lymphatic vessels or blood capillaries. See alveolus.

capsule

Protective mucilage coat surrounding certain bacteria. See bacterium for diagram.

capture-recapture technique

A method used to estimate the size of a population of a given species of animal in an area.

1. Catch and count total number and number marked on second visit, and release animals.

2. Catch, mark, count and release animals on first visit. Use following formula:

3. Size of population = $\dfrac{\text{1st count} \times \text{total 2nd count}}{\text{number marked on 2nd count}}$

carbohydrate

Composed of elements carbon, hydrogen and oxygen. Called saccharides – 3 types:

*   Monosaccharides: contain 6 carbon atoms or less in their chemical formula e.g. 5C = pentose sugar (e.g. ribose $C_5H_{10}O_5$, deoxyribose $C_5H_{10}O_4$) 6C = hexose sugar (e.g. glucose $C_6H_{12}O_6$, fructose $C_6H_{12}O_6$, galactose $C_6H_{12}O_6$)

*   Disaccharides: composed of two monosaccharide units (e.g. sucrose = glucose + fructose, maltose = glucose + glucose)

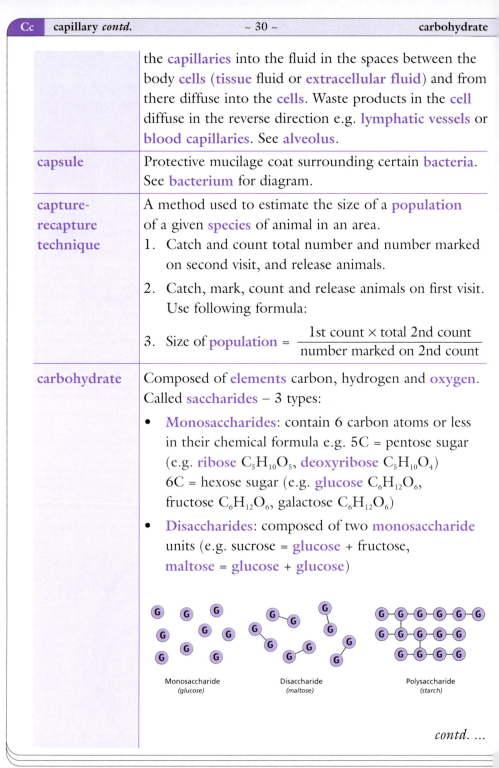

Monosaccharide   Disaccharide   Polysaccharide
(glucose)   (maltose)   (starch)

contd. ...

- **Polysaccharides**: composed of many sugar units e.g.
  - **starch** – plant storage **carbohydrate**
  - **glycogen** – animal storage **carbohydrate**. Stored in **liver**, **muscle** and **brain**.
- **Cellulose** is stored in **cell wall** of plants only.

carbon cycle	The way in which carbon circulates in nature, between the air, plants, animals etc.

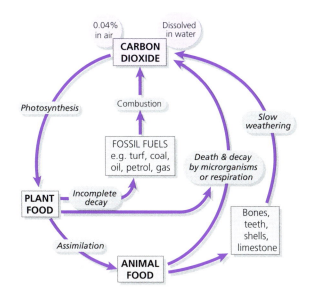

carcinogen	Any substance that will cause a normal **cell** to become a **cancer cell** e.g. formaldehyde, carbon tetrachloride, radiation, smoke from cigarettes, etc.

cardiac cycle	This is the sequence of events that produce one **heartbeat**: 1. **Tricuspid** and **bicuspid valves** closed so both **atria** fill. When **atria** are full, **blood** flow to **heart** stops. 2. The **atria** contract (atrial **systole**) and force **blood** into **ventricles**. 3. Then **ventricles** contract (ventricular **systole**) and **atria** relax (atrial **diastole**) forcing **blood** out the **pulmonary artery** and **aorta** and closing the **tricuspid** and **bicuspid** valves with a lub sound. 4. **Ventricles** relax (ventricular **diastole**) closing the semi-lunar valves with a dub sound. 5. **Valves** prevent backflow of **blood**.	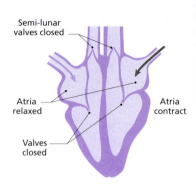 **Atrial Diastole** *(atria fill with blood)* 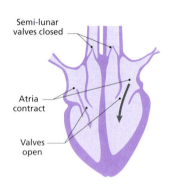 **Atrial Systole** *(blood pumped to ventricles)* 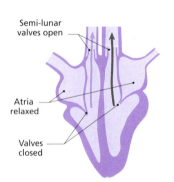 **Ventricular Systole** *(blood pumped out of heart)*
cardiac muscle	**Involuntary**, striated **muscle** of the **heart**. Does not fatigue.	

cardiovascular system	Circulatory system of the human made up of the heart and the blood vessels that transport blood to and from the heart.
carnivore	An animal or plant that feeds exclusively or mainly on animal flesh. A meat eater e.g. fox, dog, ladybird, Venus fly-trap. Compare herbivore and omnivore.
carpal	One of eight bones in the wrist. See the arm for diagram.
carpel	Female reproductive part of a flower, consisting of stigma, style and ovary. See flower for diagram.
carrier	An individual or organism which transmits pathogenic organisms e.g. humans, flies, etc. See vector. Or In genetics, a heterozygous individual (e.g. Aa) which can transmit its recessive abnormal gene (e.g. for albinism) to its offspring but will not be affected by the gene itself.
cartilage	Tough, elastic tissue (not as hard as bone) composed of protein fibres. Found on ends of long bones, nose and ear. Consists of living cells scattered throughout a non-living ground matrix. Prevents friction between bones and acts as a shock absorber. See bone for diagram.
Casparian strip	Ring of cells of endodermis, which are impermeable to water. Water and solutes can only pass through endodermis via the cytoplasm of passage cells.
catabolism	A chemical reaction which breaks down a large molecule into simpler ones using enzymes e.g. respiration. Catabolic reactions release energy. See metabolism and anabolism.
catalase	Enzyme found in the tissues of plants and animals (e.g. celery and liver). Breaks down hydrogen peroxide (H_2O_2), which is poisonous, to water and oxygen.
catalyst	Substance used to speed up or slow down a chemical reaction but does not get used up in the reaction.

cell	Basic/smallest structural unit of living things capable of functioning independently. Contains protoplasm and is surrounded by a cell membrane.

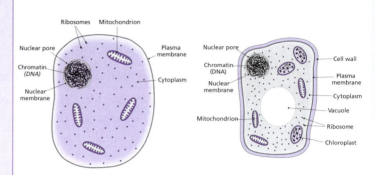

cell body	Controls activities of neurons. Contains Nissl's granules (composed of RNA and used in protein synthesis), nucleus and mitochondria and produces a neurotransmitter. See neuron for diagram.
cell cycle	The sequence of steps performed by a cell in order to replicate itself (replication), from one division by mitosis to the next. The cycle consists of a growth period during which the cell increases in size followed by the replication of its DNA and the division of the cell (cell division).

cell division | The process by which new cells are made from existing cells. See mitosis and meiosis.

cell membrane | Also called the plasma membrane or plasmalemma. Double layer, living. It encloses the protoplasm. Capable of growth, is flexible and has a lipo-protein nature. Porous, semi-permeable. Retains cell contents. Allows diffusion, osmosis, and active transport. Receives stimuli e.g. hormones. Is the site of metabolic reactions, and elimination of waste. Has a protective function (recognises foreign particles). See cell for diagram.

cell plate | In a plant cell during telophase of mitosis, a cell plate is formed to divide the cell in two. It later forms the middle lamella.

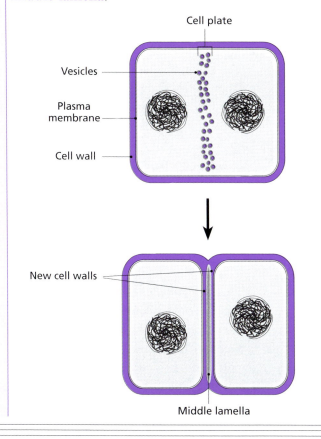

Cell plate

Vesicles

Plasma membrane

Cell wall

New cell walls

Middle lamella

cell theory	Theory put forward in 1838–39 by Schleiden and Schwann, which stated that all plants and animals are made up of cells and all cells arise from pre-existing cells through cell division.
cell wall	Only in plants. Non-living, rigid, and fully permeable. Found outside the cell membrane, made of cellulose by the cytoplasm. Gives strength and support. See cell for diagram.
cellulose	Complex carbohydrate (polysaccharide) forming the cell walls of plant cells. See fibre.
central nervous system (CNS)	A mass of nerve cells (nervous tissue) which co-ordinates the activities of an animal. The CNS of vertebrates is composed of the brain and spinal cord. See nervous system for diagram.
centriole	Found in many non-dividing cells. During mitosis it doubles and moves to opposite sides of the cell where it forms the poles of the spindle fibres.
centromere	The point of attachment of the spindle to the chromatid. Pairs of chromatids are held together by the centromere. Used to attach to spindle during metaphase of cell divisions. See chromosome for diagram.
cerebellum	That part of the human brain lying beside the medulla oblongata. It co-ordinates voluntary muscle activity and balance in the body. See the brain for diagram.
cerebral hemisphere	One of two halves of the cerebrum. See the brain for diagram.
cerebrospinal fluid	Fluid of the brain and spinal cord.
cerebrum	That part of the human brain found above the pons and medulla oblongata. It is made up of two halves, the cerebral hemispheres, which are connected together by a band of nerve fibres – the corpus callosum. It is responsible for movement, sensations, learning, memory, intelligence and emotions. See the brain for diagram.

cervix	Neck or neck-like structure e.g. neck of womb (uterus). Ring of muscle to retain developing embryo. See the female reproductive system for diagram.
CFC	Chlorofluorocarbons. Chemical substances used in refrigerators, air conditioners, aerosol cans and solvents. They drift to the upper atmosphere and break up, releasing chlorine which reacts with ozone, thus eroding the ozone layer.
characteristic	Distinctive trait, mark, or feature possessed by organism.
characteristics of life	Distinctive features or traits possessed or shared by all living organisms. These are sensitivity (or responsiveness), nutrition, organisation, reproduction and excretion i.e. SNORE.
chemoreceptors	Sensory receptors that respond to the presence of chemicals e.g. receptors in the tongue (taste) and nose (smell). See receptor.
chemosynthesis	Formation of carbohydrates from inorganic compounds without sunlight. See chemosynthetic bacteria.
chemosynthetic bacteria	e.g. nitrifying bacteria. Autotrophic bacteria (organisms) that get their energy from the oxidation of inorganic compounds (e.g. nitrogen compounds and not from light (photosynthesis).
chemotropism	The growth response of a plant to chemicals e.g. fertilisers.
chlorophyll	Green pigment found in plants necessary for trapping light energy for photosynthesis. Contains the element magnesium. Made in chloroplasts.

chloroplast	A specialised cell organelle (plastid) containing chlorophyll. Chloroplasts are green in colour and functions during photosynthesis. They are composed of grana (used during the light stage) and stroma (used during the dark stage) of photosynthesis.

cholesterol	Fatty substance synthesised in the liver from certain animal fats. High levels of cholesterol are associated with thickening of blood vessels (arteriosclerosis) and may lead to a heart attack or stroke. Involved in hardening of the arteries. See heart attack.
chordae tendinae	Non-elastic strands of tissue found in heart joining the bicuspid (mitral) and tricuspid valves to the walls of the ventricles. Their function is to prevent the valves blowing inside out when the ventricles contract.
chorion	Outermost membrane surrounding the embryo, involved in placenta formation. See embryo for diagram.
choroid	Dark pigmented layer inside the sclera of the eye. Prevents light reflection in the eye. Contains blood vessels. See the eye for diagram.
chromatid	During cell divisions each chromosome replicates to produce two identical chromatids (sister chromatids) joined together at the centromere i.e. it is a duplicated chromosome or a chromosome during replication. The chromatids separate during anaphase and are known as chromosomes.

chromosome   Rod/thread-like structure composed of DNA and protein, contains the genetic information (genes) which is passed from one generation of cells or organisms to the next. Occur in pairs in most plant and animal cell nuclei.

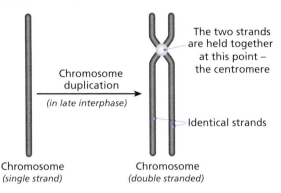

Chromosome duplication
(in late interphase)

The two strands are held together at this point – the centromere

Identical strands

Chromosome
(single strand)

Chromosome
(double stranded)

chromosome mutation

There are two types:

1. Changes in structure of chromosome i.e. in the position, arrangement or number of genes present.

2. Changes in number of chromosomes:
   (i) Addition or loss of one or more chromosomes, caused by failure of chromosomes to separate during cell division e.g. Trisomy 21 (Down's syndrome or mongolism).
   (ii) Polyploidy: the addition of a set of chromosomes. n = haploid, 2n = diploid, 3n = triploid, etc. Polyploid plants grow vigorously, give greater yields and have good vegetative reproduction.

chronic   Relates to pain/disease. Lasting or lingering. Compare acute.

cilia   (Singular = cilium.) Tiny, hair-like cytoplasmic extensions of a cell which beat rhythmically. In lower animals they are for locomotion; in higher forms, for propulsion of substances e.g. mucus.

ciliary body	Front portion of the choroid in the eye which contains the ciliary muscles. See the eye for diagram.
ciliary muscles	Circular muscles that alter the shape of the lens of the eye and are used for focusing the image onto the retina. See the eye for diagram.
ciliated epithelium	Epithelial tissue (cells) which contain minute hair-like cytoplasmic extensions (processes) which beat rhythmically e.g. the cells lining the trachea.
circulatory system	The heart, which pumps the blood, and the blood vessels, which convey the blood around the body. In humans it is composed of the systems for pulmonary circulation and systemic circulation. See pulmonary circulation for diagram.
classification	Taxonomy. The arrangement of animals and plants into groups based on common characteristics. The groupings are: kingdom, phylum, class, order, family, genus and species.
clavicle	The collar bone, from front of neck to shoulder. See the human skeleton for diagram.
cleavage	The dividing of the cytoplasm of the cell into two during telophase of mitosis.
cleavage furrow	Constriction of the cell membrane which results in dividing the cytoplasm of animal cells at the equator. The cell divides as the furrow deepens.
climatic factors	Elements of the climate (weather) that influence the life and distribution of the organisms that live in a particular environment e.g. rainfall, humidity, temperature, light intensity and day length.
clone	All offspring genetically identical, produced by asexual reproduction i.e. all have come from one original parent e.g. King Edward potato. A clone is a genetically identical copy produced by recombinant DNA technology.

closed circulation	Refers to a transport system that confines the blood to or maintains it in a collection of tubules. Compare open circulation.

Body cells

Collecting vessels — Distributing vessels

Muscular heart

coagulate	Clot, curdle or change from a liquid state to a semi-solid state.
coccus	(Plural = cocci.) A spherical bacterium. Bacteria are classified according to their shape.

Capsule

Pneumonia   Sore throat   Food poisoning and boils

cochlea	Coiled tube of inner ear sensitive to sound. Concerned with hearing. See the ear for diagram.
codon	A sequence of three nucleotides in messenger RNA that codes for a single amino acid. See triplet.
cohesion	Force with which molecules of the same substance stick to each other.
coleoptile(s)	A sheath containing the first leaf (shoot or plumule) of a germinating cereal grain.
collagen	The most abundant body protein, a primary component of mammalian hair and forms the inelastic outer layer on arteries and veins. Any disease of collagen may have long-term effects. See artery for diagram.

collecting duct	Tubule, in the medulla of the kidney, into which numerous nephrons discharge their contents (urine) for delivery to the pelvis and hence to the bladder via the ureter. Water is reabsorbed here during the formation of urine, which concentrates the urine. The quantity of water reabsorbed is controlled by the hormone ADH. See nephron for diagram.
colon	Part of the large intestine from the caecum to the rectum. See digestive system for diagram.
colonisation	When plants or animals 'take over' new ground (habitat).
colostrum	The first milk produced by a mammal after parturition. It is a thick, yellowish liquid containing proteins and antibodies, and is later replaced by breast milk.
colour blindness	Inherited condition (sex-linked recessive trait) where certain colours cannot be distinguished.
columella	A wall separating the sporangium from the sporangiophore of a reproducing *Rhizopus* (bread mould). See *Rhizopus* for diagram.
commensalism	Where one organism obtains benefit from another and leaves it completely unaffected – neither harmed nor helped e.g. the bacterial population in the intestines of humans and other animals.
community	A group of interacting populations of different species living in the same area i.e. plants and animals in a certain area (e.g. a pond) where there is interdependence of various species on each other.
compact bone	The outer layer of the shaft of the long bones. It is made up of mineral deposits surrounding a central opening. It gives strength and rigidity. It is made of living cells and is supplied by blood vessels and neurons. See bone for diagram.
companion cells	Cells in phloem associated with sieve tube cells. See phloem for diagram.

comparative anatomy	Comparing the structures of plants and animals, used as evidence that organisms share a common ancestor e.g. homologous structures, the forelimbs of many animals share the same basic structure but differ in their functions:

human → lifting and grasping	bat wing → flying
rabbit → leaping	monkey → grasping
whale fin → swimming	mole → digging
anteater → tearing	horse → running

comparative embryology	Comparing the developing embryos of vertebrates shows a similarity between certain structures e.g. the position of the brain, eyes, gill slits and tail of the fish, tortoise, rabbit, bird, chick and human. This suggests that organisms descended from common ancestors.
competition	The struggle between organisms to obtain a sufficient supply of a resource of limited quantity i.e. animals compete with each other for food, shelter and space, and plants compete for space, light, water and minerals. See contest competition and scramble competition.
complement proteins	A number of blood proteins that make holes in the surface of bacteria and viruses, causing them to burst.
compound	A substance formed when two or more elements are joined together chemically e.g. CO_2, H_2O, $C_6H_{12}O_6$, etc.
concentration	The amount or number of molecules of a solvent, which is dissolved in a solute.
concentration gradient	The difference in the concentration/number of molecules between two different areas. The greater the difference, the faster the rate of movement of molecules will be e.g. diffusion, active transport, etc.

conception	Being conceived. The very beginning of a pregnancy when an egg is fertilised by a sperm, develops into a blastocyst and is implanted in the wall of the womb (uterus) i.e. conception = fertilisation + implantation.
condensation	In biology it is the joining together of molecules with the production/elimination of water e.g. glucose + glucose $\rightarrow$ maltose + water $C_6H_{12}O_6 + C_6H_{12}O_6 \rightarrow C_{12}H_{22}O_{11} + H_2O$
condom	A rubber-like sheath worn: • by males over the erect penis • by females inserted into the vagina during sexual intercourse to prevent sperm entering the vagina. (The female condom is a different type of condom.) A condom is a barrier method of contraception.
cone(s)	(In eye.) Cells in the fovea of the retina of eye stimulated by bright light. There are three types of cone. Each type responds to a different colour (red, green, and blue) giving colour vision. See retina for diagram.
connective tissue	Tissue used for attaching organs together or for protection e.g. vertebrate tissue consisting of few cells with collagen or elastic fibres between them. Is the main tissue of bone, tendons, ligaments, cartilage, etc.
conservation	This is the protection and wise management of natural resources and the environment. The benefits are that: • existing environments are maintained, • endangered species are preserved for reproduction, • the balance of nature is maintained, • pollution and its effects are reduced.

consumer	A heterotrophic organism (heterotroph) that feeds on other organisms. It cannot make its own food. Two types: 1. Primary consumer eats plants (i.e. herbivores). 2. Secondary consumer eats herbivores (i.e. carnivores). See also omnivores and detrivore.
contest competition	Involves an active physical confrontation between two organisms e.g. two dogs fighting over a bone. One may have stronger muscles and sharper teeth and so will win the bone.
continuity	Being passed on from one generation to the next. This requires organisation, nutrition, behaviour, growth, synthesis, and reproduction.
continuity of life	The ability of organisms to exist from one generation to the next.
continuous flow processing	The organisms being grown are maintained in a particular phase of the growth curve. Compare batch processing. See growth curve for diagram.
contraception	The act of preventing the fertilisation of an egg and subsequent pregnancy. Different methods of birth control include: • natural: abstaining from sexual intercourse during the fertile period • mechanical: use of a barrier e.g. condom, to prevent sperm and egg meeting or use of an IUD (intrauterine device) to prevent implantation • chemical: use of the pill to prevent ovulation • surgical: tubal ligation in females and vasectomy in males.

contractile vacuole	Structure possessed by *Amoeba* (a unicellular aquatic organism) necessary to prevent cell bursting since water enters the cell by osmosis and, together with water from respiration, accumulates in the cell. The excess water is expelled by contractile vacuole. Marine *Amoeba* are isotonic with their surroundings, therefore no contractile vacuoles required. See *Amoeba* for diagram.
control	Used in experiments as a standard against which your results can be compared in order to check their validity.
corm	Modified stem – short, vertical, swollen. Terminal bud produces leaves and flowers above ground. New corm formed on top of old e.g. crocus, gladioli.
cornea	Transparent part of sclera at the front of the eye. Allows light to enter eye. See the eye for diagram.
coronary blood vessels	These are the blood vessels which surround the heart and supply it with blood.
coronary thrombosis	The development of a blood clot in a coronary artery (see coronary blood vessels), causing the artery to become blocked and results in the section of the heart supplied by that artery becoming starved of blood and deprived of oxygen, leading to a possible heart attack.
corpus callosum	Tightly bundled nerve fibres that join the two cerebral hemispheres of the brain. See brain for diagram.
corpus luteum	A yellow body that develops in the ovary after the rupture of the Graafian follicle (ovulation). It is a source of progesterone, persists if pregnancy has occurred.
corpuscle(s)	A cell, especially one floating in fluid such as blood (e.g. red blood cell) or saliva or a sensory nerve ending enclosed in a capsule.
cortex (of kidney)	The outermost layer of the kidney. See the kidney for diagram.

cortex (of plant)	Tissue of the stem or root that lies outside the vascular tissue. 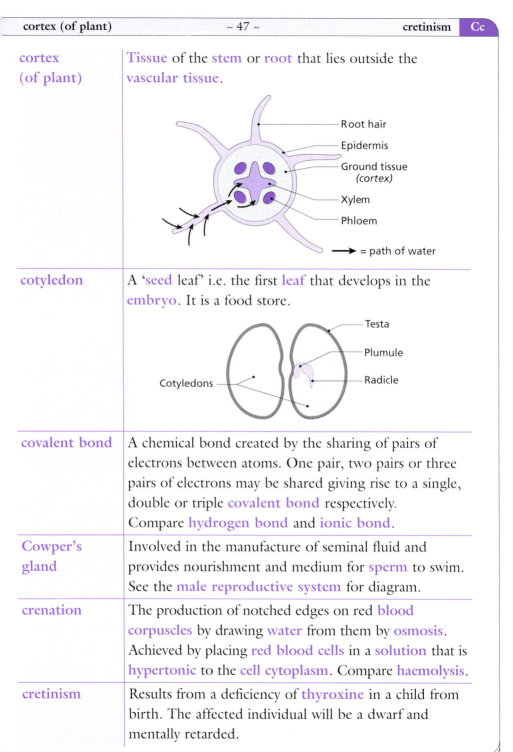
cotyledon	A 'seed leaf' i.e. the first leaf that develops in the embryo. It is a food store.
covalent bond	A chemical bond created by the sharing of pairs of electrons between atoms. One pair, two pairs or three pairs of electrons may be shared giving rise to a single, double or triple covalent bond respectively. Compare hydrogen bond and ionic bond.
Cowper's gland	Involved in the manufacture of seminal fluid and provides nourishment and medium for sperm to swim. See the male reproductive system for diagram.
crenation	The production of notched edges on red blood corpuscles by drawing water from them by osmosis. Achieved by placing red blood cells in a solution that is hypertonic to the cell cytoplasm. Compare haemolysis.
cretinism	Results from a deficiency of thyroxine in a child from birth. The affected individual will be a dwarf and mentally retarded.

crista(e)	A fold on the inner membrane of the mitochondrion. Creates a large surface area. Site of oxidative phosphorylation i.e. the hydrogen carrier system in cellular respiration.
cross-fertilise	(Animal or plant.) The union of haploid male and female gametes produced by different members of the same species, or from a member of a different species. Compare self-fertilise.
cross pollination	The transfer of pollen from the anther of the stamen of one flower to the stigma of the carpel of another flower on a different plant of the same species.
culture	Cultivate (grow) micro-organisms or tissue cells on a special medium. Or A mass of organisms or cells grown on such a medium.
cuticle	Non-cellular waxy layer covering the epidermis of a leaf. Prevents excess loss of water (by evaporation) from leaf.

Cuticle

Upper epidermis (dermal tissue)

Palisade mesophyll

Vascular bundle (vascular tissue)

Ground tissue

Spongy mesophyll

Air space

Lower epidermis (dermal tissue)

Stoma   Guard cells

cyclic photophos-phorylation	Part of the light stage/phase of photosynthesis involved with the production of energy (ATP). Summarised as follows: • light strikes a chlorophyll molecule and excites an electron, • electron emitted and taken up by an electron acceptor.

- Electron then passed through a series of carriers. As electron is passed from one carrier to the next, energy is removed from it for the conversion of ADP to ATP.
- The last carrier hands the electron back to the chlorophyll molecule.

Note: The electron returned to the chlorophyll molecule is the same one that was emitted, hence 'cyclic'.

cystic fibrosis	One of the most common genetic disorder diseases in children. *Caused by*: a disorder in a gene on chromosome seven, and is a recessive condition, so both parents may be carriers without having the disease. *Effects*: it causes the production of mucus that clogs the airways of the lungs and the ducts of the pancreas and other secretory glands. *Symptoms*: include failure to gain weight with frequent chest infections and loose, pale stools. Sodium chloride concentration is increased in sweat. *Treatment*: consists of taking pancreatic enzymes, as well as preventing and treating respiratory infections. Heart and lung transplants, as well as genetic manipulation, may provide the answer for future. It is possible that in the future cystic fibrosis screening may become a routine antenatal investigation. At the moment, genetic counselling is offered to those whose children might be affected.
cytokinins	Plant growth hormones, which increase the rate of cell division and inhibit aging of green tissues in plants.
cytoplasm	The contents of a cell excluding the nucleus. cell = cell membrane + protoplasm = cell membrane + cytoplasm + nucleus ⇒ protoplasm = cytoplasm + nucleus See the cell for diagram.
cytosine	Nitrogenous base of the nucleic acids, DNA and RNA – a pyrimidine.
cytosol	The cytoplasm minus the cellular organelles.

Dd

dark stage (of photosynthesis)	Takes place in the stroma of the chloroplasts. Is a light independent stage i.e. it can take place in the presence or absence of light. The function of the dark stage is the reduction of carbon dioxide i.e. adding hydrogen to CO_2 to form a carbohydrate – glucose. It is an endergonic process. Energy (ATP) produced during the light stage is used here; hydrogen, supplied by $NADPH_2$, is used to reduce CO_2.
deamination	Removal of the amino group ($^-NH_2$) from an amino acid. This is done to excess amino acids in the liver where the amino group is converted to ammonia and then urea (which is excreted by the kidneys).
death phase	The fourth phase in a growth curve. Numbers decline due to increased competition for food and space and the build-up of toxic wastes. See growth curve.
deciduous	Plant that sheds its leaves and has a period of dormancy annually e.g. oak, ash, etc.
decomposer	An organism which feeds on and breaks down the dead remains of organisms (detritus) and excreta into simpler substances e.g. bacteria, fungi etc.
defence against disease	Methods used by the body to resist infection i.e. to gain immunity. Different types are natural immunity, acquired immunity, active immunity and passive immunity.
dehiscence	The bursting open or splitting of a plant structure (e.g. capsule, seed pod, fruit or anther) to release its contents.
denaturation	Changes in the shape of an enzyme (protein) due to extreme conditions, such as high temperature, unsuitable pH, or the presence of certain chemicals. If these changes are permanent they destroy the activity of the enzyme and the enzyme is denatured.

dendrite	A short branched process (extension/fibre) of a nerve cell (neuron), which receives impulses and transmits them towards the cell body. See neuron for diagram.
dendron	A process (a piece sticking out) of the cell body of a neuron, which transmits impulses towards the cell body. See neuron for diagram.
denitrification	Removal of nitrates and nitrites from the soil by converting them to ammonia and then nitrogen gas. This is done by denitrifying bacteria e.g. *thiobacillus denitrificans*. See nitrogen cycle.
dental formula	A method used to indicate the number of different types of teeth found in the upper and lower jaws of one side of the mouth, in mammals. e.g. human $i\frac{2}{2}$, $c\frac{1}{1}$, $p\frac{2}{2}$, $m\frac{3}{3}$ $\frac{\text{upper jaw}}{\text{lower jaw}}$ Total = 32 Note: i = incisor; c = canine; p = premolar; m = molar
dentition	The shape, number and arrangement of teeth of a species or individual mammal. Varies with diet.
deoxyribose	$C_5H_{10}O_4$. Ribose sugar ($C_5H_{10}O_5$) with one atom of oxygen removed. Five-carbon sugar found in the nucleotides of DNA.
deplasmolysis	The opposite of plasmolysis. The cell membrane moves towards the cell wall as a result of water being drawn into the cell by osmosis. The cell gradually becomes turgid. Achieved by placing a plant cell in a solution that is hypotonic to the cell sap. See turgor for diagram.
dermis	Found beneath the epidermis. Connective tissue, elastic fibres. Contains sensory, excretory organs etc. See the skin for diagram. Or Outer layer or covering of plant. See vascular bundle for diagram.

descending tubule	Forming that part of the nephron just before the loop of Henle through which the glomerular filtrate flows downwards. Permeable to water only. See nephron for diagram.
detritus	Dead organisms.
detritus food chain	Food chain composed of animals and plants, which cannot photosynthesise and so get their food by feeding on dead plant material. Or Food chain that begins with dead organic matter and animal waste (detritus): e.g. 1. detritus → edible crab → seagull e.g. 2. fallen leaves → earthworms → blackbirds → hawks
detrivore	An organism that gets its nutrition from the detritus in an ecosystem. See decomposer.
diabetes insipidus	Excessive production of watery (dilute) urine – caused by insufficient production of ADH (vasopressin) by the pituitary gland.
diabetes mellitus	Sugar in urine and excessive urine production caused by an inadequate production of insulin by the pancreas.
diaphragm	A muscular sheet or membrane separating the thorax and abdomen. Situated below the lungs and above the liver, stomach and spleen. It is dome-shaped at rest. See inspiration and expiration. See the respiratory system for diagram.
diastole	Relaxation phase of cardiac cycle during which the chambers fill with blood. Compare systole.

| dicotyledon (dicot) | Plants that may be woody or herbaceous, have an embryo with two cotyledons (a non-endospermic seed it stores food and passes it to the embryo). The same can be said for flower parts in units of four or five, leaves with netted (reticulate) venation and vascular bundles arranged in a circle. Dicot plants also have tap roots e.g. broad bean. Compare monocot (monocotyledon). |

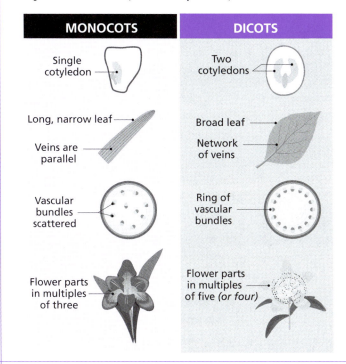

| differentiation (zone) | Area in root and shoot tips where simple cells become modified and specialised to perform specific tasks (e.g. xylem, phloem, etc.) depending on where they are located. Differentiation leads to the formation of: |

(a) dermal tissue which forms the protective covering on the outside of plants

(b) ground tissue which forms the bulk of and fills the interior of the plant

(c) vascular tissue i.e. xylem and phloem which form the transport tissues in plants.
See root cap for diagram.

diffusion	The movement of **solute** from a region of high **solute concentration** to a region of lower **solute concentration**. No **permeable** or **semi-permeable membrane** is necessary for **diffusion** to occur. No **energy** is used by the **cell** in **diffusion** i.e. it is a passive process. Examples of **diffusion**: **gaseous exchange** in **alveoli**; **absorption** through **villi** of **small intestine**. Compare **osmosis**.
digestion	The physical (chewing = mastication) and chemical process by which large particles and **molecules** of food are broken down into simpler, soluble, absorbable, usable forms.
digestive system	Group of **organs** concerned with **digestion**, **stomach**, **pancreas**, **small intestine**, etc.

Mouth

Salivary gland

Salivary gland

Pharynx

Oesophagus

Liver

Gall bladder

Stomach

Duodenum

Pancreas

Colon

Caecum

Ileum

Appendix

Rectum

Anus

digits	Another name for the fingers or toes. See arm for diagram.
dihybrid cross	A genetic cross which examines the transmission of two traits together e.g. shape and colour of seed (green and wrinkled). Compare monohybrid cross.
diploid (2n)	Every organism has a fixed number of chromosomes. If they occur in pairs the organism is said to be diploid (2n) i.e. two sets of chromosomes – one set received from the father and one set from the mother at fertilisation e.g. human 2n = 46, mouse 2n = 40, fruit fly 2n = 8.

Homologous pairs   Maternal chromosomes   Paternal chromosomes

2n = 4   2n = 6   2n = 8

	Or Double the number of chromosomes found in gametes. Compare haploid.
disaccharide(s)	A carbohydrate composed of two monosaccharide units (e.g. sucrose = glucose + fructose, maltose = glucose + glucose). General chemical formula = $C_x(H_2O)_y$, where x is approximately equal to y e.g. sucrose = $C_{12}H_{22}O_{11}$.
disease	Unhealthy condition in which an organism or part of an organism is affected by some factor which interferes with normal growth, development or metabolism of its organs or tissues.
disinfectant	An agent which kills micro-organisms (but not the spores) on contact but does not prevent new growth developing. Use is restricted to inanimate matter because of its toxic properties.

dispersal	The carrying of seed or fruit away from the parent plant, necessary to avoid overcrowding, minimise competition and encourage colonisation. This increases the chances of survival of the seed. Agents of seed dispersal include wind, water, animal and self-dispersal.

Dandelion — Hairs, Fruit

Sycamore — Wings, Fruits

distal convoluted tubule	One of two highly coiled tubules in the nephron of the kidney, this one situated furthest away from the Bowman's capsule and close to the collecting duct. Water is reabsorbed here during the formation of urine. See nephron for diagram.
diuresis	An increase in production of urine by the kidneys, either from increased fluid intake, alcohol (inhibits ADH production), renal disease, diabetes insipidus or diuretic therapy.
diurnal	Refers to an organism that is active during the day. Compare nocturnal.
diversity	The different types of organisms that are found in a community.
Dixon-Joly theory	Most acceptable theory of water movement through a plant. Based on cohesive forces (cohesion) of water i.e. the force of attraction between water molecules is great enough to maintain a continuous column of water. Explained as follows:
	1. In the leaf, spongy mesophyll cells lose water by evaporation to the air spaces and then to the atmosphere.
	2. These cells lose turgidity and draw more water from surrounding cells to try and maintain turgidity.

3. This results in the formation of a suction pressure from air spaces to xylem vessels.
4. Cells closest to the xylem vessels absorb water from the xylem by osmosis.
5. This creates a pull on the water in the xylem and draws the water upwards.
6. The water is held at the top of the xylem by the adhesion of water molecules to the xylem wall.

DNA	Substance found in cell nuclei in the chromosomes. Regulates protein synthesis and is the main molecule of genes.

DNA ligase	An enzyme that joins two DNA fragments from different sources together to form a recombinant DNA molecule.
DNA polymerase	Enzyme that forms and repairs DNA.
DNA profile	A picture (auto radiograph) of the pieces of DNA produced when an organism's DNA is broken up using specific enzymes and then sorted by size on a gel. The stages involved are: 1. Cells are broken down to release DNA. 2. DNA strands are cut into fragments using restriction enzymes. 3. Fragments are separated on the basis of size using gel electrophoresis. 4. The pattern of fragment distribution is analysed.

dominance	(Genetics): characteristic, trait, or gene that expresses itself in offspring, even when the corresponding opposite one (recessive) is also inherited. e.g. Tt = tall, T is dominant. Dominant gene = gene that is expressed in the heterozygous condition. Or (Ecology): in a habitat an organism or group of organisms which, because of their size or numbers, determine the character of the habitat.
dopamine	A neurotransmitter. Lack of this causes Parkinson's disease.
dormancy	A period of rest, inactivity or non-vegetative state before growth, during which the rate of metabolism is reduced e.g. in buds, seeds and spores. Seeds will not germinate during this time, even if given ideal conditions, because other requirements may be necessary before germination can occur. For example: • seed coat (testa) too hard, must wait for it to be softened by nature • cold conditions necessary, ensures springtime germination • desert plants have chemical inhibitors in seed that must be washed out by heavy rain. Ensures water for further growth • embryo not mature. Seed not 'ripe' so must wait.
dorsal	Of, on or near the upper surface or back or posterior of a body.
dorsal root	Pairs of projections arise from the spinal cord. These are called spinal roots, of which there are two types: • the dorsal root, which has a swelling, carries sensory neurons into the spinal cord • the ventral root, which carries motor neurons away from the spinal cord. See reflex arc for diagram.

'double-blind' testing	e.g. during trials to examine the effectiveness of a new pill, two sample groups are taken. One group gets the pill under test and the other a placebo. All pills are the same size, colour, taste, etc., but coded differently. When the trials are over, it is possible to distinguish between the control and experimental groups. This 'double blind' testing method avoids bias during the trials.
double circulatory system	Humans have a double circulatory system i.e. blood is pumped from the heart to the lungs and back to the heart (this is pulmonary circulation); and from the heart to the body and back to the heart (this is systemic circulation).
double fertilisation	This is a characteristic of angiosperms. The pollen tube carries two sperm nuclei to the female embryo sac in the ovule. One sperm nucleus fuses with the egg cell and gives rise to a diploid embryo. The other sperm nucleus fuses withthe two polar nuclei to form a triploid endosperm nucleus.
double helix	Refers to the shape of the DNA molecule. It is like a twisted rope ladder, each upright forms a spiral (helix). Two uprights – two spirals = double helix.

Base pair

dry weight	The weight of tissue after all water has been removed.
ductless gland	See endocrine gland.
duodenum	The first portion of the small intestine immediately below the stomach, approximately 30 cm long. See the digestive system and bile duct for diagram.
dynamic equilibrium	Where changes in the population of species are in an unchanging state of balance, without loss or gain of species and the population of each remaining relatively constant. Any disturbance in population size tends to return to the equilibrium.
dysmenorrhoea	Medical term for menstruation with severe cramps, diarrhoea, backache, nausea and vomiting. It may be a symptom of endometriosis or fibroids in older women.

Ee

ear	Organ of hearing and balance.

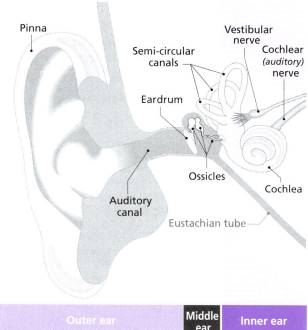

Pinna

Semi-circular canals

Vestibular nerve

Cochlear *(auditory)* nerve

Eardrum

Ossicles

Cochlea

Auditory canal

Eustachian tube

Outer ear | **Middle ear** | **Inner ear**

ear ossicles	See ossicles.
eardrum	Also tympanum, a membrane separating the outer ear and the middle ear. Converts sound waves to vibrations. See the ear for diagram.
ecological pyramids	A way of comparing different communities of an ecosystem in order of different trophic levels (feeding levels). See pyramid of numbers.
ecology	The study of how living things relate to each other and to their environment.

ecosystem	A community of living organisms interacting with one another and their environment within a particular area e.g. woodland, hedgerow, sea shore, tree etc. The earth itself is a true ecosystem as no part of it is completely isolated from the rest. Ecosystem = communities + environment.
ectoderm	Outer of the three primary germ layers. Gives rise to skin, hair, and nails. See germ layer for diagram.
ectoparasite	A parasite that lives on the surface of an organism e.g. flea on a dog. Compare endoparasite.
ectotherm	Animal whose temperature changes with that of its surroundings i.e. a cold-blooded animal e.g. fish, frog, reptile. It gains heat by moving into a warmer area and loses heat by moving into a colder area. Compare endotherm.
edaphic factors	The physical, chemical and biological characteristics of the soil that influence the community. The major edaphic factors are: available (soil) water, mineral content, pH, humus, soil texture and structure.
effector	A muscle or a gland that responds when stimulated by a nerve impulse.
efferent neuron	A nerve carrying messages from the central nervous system (CNS) to effectors. Cell body located at end of axon, inside CNS e.g. motor neurons are efferent neurons. See neuron and reflex arc for diagram.
egestion	Elimination of faeces from the body i.e. undigested or undigestible and unwanted material. It is under the control of the anal sphincter muscle. This is not excretion.
egg	Sphere shaped body produced by female, containing germ of new individual. Capable of developing into new individual when fertilised by male sperm.

After ovulation the human egg can be fertilised for a period of about 48 hours.

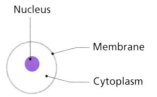

Nucleus

Membrane

Cytoplasm

electron microscope	A microscope that uses electrons instead of light to magnify objects. Capable of x 300,000 and higher magnifications. Two types: 1. SEM = Scanning Electron Microscope – electrons reflected from surface of object and photographed. 2. TEM = Transmitting Electron Microscope – electrons passed through object and photographed.
element	A substance consisting of only one type of atom. The six most important elements to living things are carbon (C), hydrogen (H), oxygen (O), nitrogen (N), phosphorus (P) and sulphur (S).
elongation (zone)	Area in root and shoot tips immediately behind the meristematic zones where simple cells undergo a lengthening process prior to differentiation.
embryo	Fertilised egg or immature animal in uterus (up to eight weeks in humans).

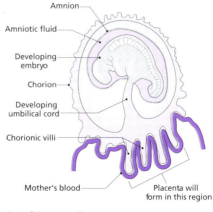

Amnion

Amniotic fluid

Developing embryo

Chorion

Developing umbilical cord

Chorionic villi

Mother's blood

Placenta will form in this region

Or

Plant contained in seed.

embryo at 3 months	At the end of the third month of development, the eyes are low and widely spaced. Bone tissue appears, grows and cartilage is replaced. Nervous system and muscles become co-ordinated, arms and legs begin to move. Sex organs differentiate, male or female is obvious in the 12th week. Foetus sucks its thumb, kicks, baby teeth begin to grow. Breathes amniotic fluid in and out, passes urine and faeces into the amniotic fluid. During the remaining months the baby grows.
embryo sac	Large oval cell in nucellus of ovule. Produces an ovum at micropylar end (see micropyle) and two polar (or endosperm) nuclei in the centre.

This diploid cell divides by meiosis *(once)* and mitosis *(three times)*

Two polar nuclei (n)

Ovule
Ovary

Embryo sac

Egg cell (n)

emulsify	Change into a smooth creamy liquid.
endergonic	Refers to chemical reactions that require an energy input to start them, e.g. photosynethesis. Compare exergonic.
endocrine	Refers to secretions made directly into the blood. See endocrine gland.

endocrine gland(s)	**Ductless gland**. Secretions (**hormones**) delivered directly into bloodstream by the following **glands**: **pituitary**, **thyroid**, **parathyroid**, **pancreas**, **adrenal**, **ovary** and **testis**. Compare **exocrine gland**.

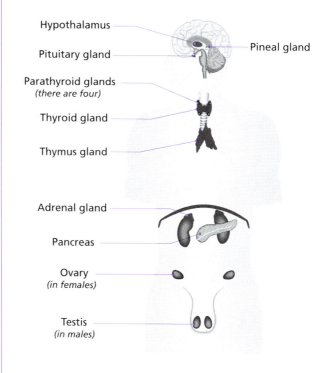

Hypothalamus

Pituitary gland

Pineal gland

Parathyroid glands
(there are four)

Thyroid gland

Thymus gland

Adrenal gland

Pancreas

Ovary
(in females)

Testis
(in males)

endocrine system	All the **endocrine glands** responsible for coordinating the widespread activity of internal **organs** at a slower response rate than the **nervous system**. See **endocrine gland** for diagram.
endoderm	The innermost of the primary **germ layers**. Gives rise to the linings of the **alimentary canal**, **trachea** and **bronchi**. See **germ layer** for diagram.

endometriosis	A condition where fragments of the uterus lining migrate to other parts of the pelvic cavity, and stick to the outside of various organs e.g. ovaries, bladder, uterus, vagina. They continue to respond to the menstrual cycle hormones and bleed each month. They cause pain during menstruation and sexual intercourse. The blood cannot escape and causes painful cysts to grow on the pelvic organs. May be due to a hormone imbalance or a weakness in the immune system that allows the fragments to become attached.
endometrium	Mucus membrane that lines the uterus (womb), undergoes cyclical changes during the menstrual cycle. See female reproductive system for diagram.
endoparasite	A parasite living inside a host e.g. liverfluke. Compare ectoparasite.
endoskeleton	Found inside body. Made of bones e.g. fox, human. Compare exoskeleton.
endosperm	Nutritive tissue found within the embryo sac of a seed plant. Arises from the union of a generative nucleus (a sperm cell) and two polar nuclei to form a triploid endosperm nucleus (3n). This nucleus divides repeatedly by mitosis and forms the endosperm, which acts as a good store. 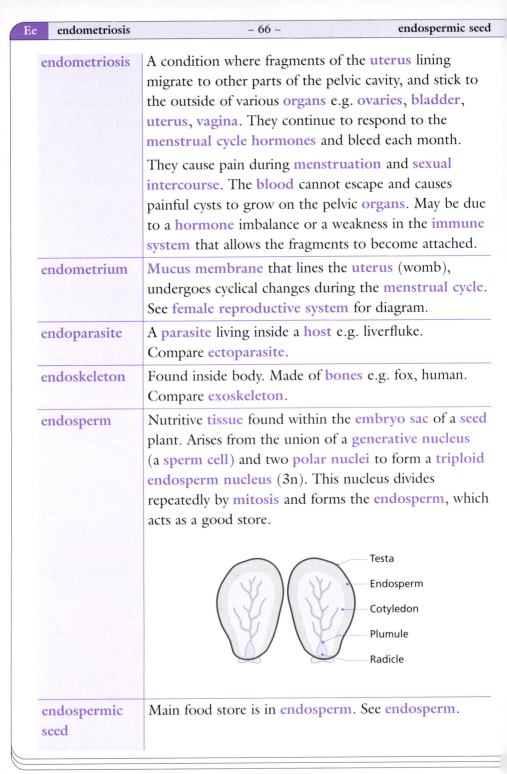 Testa Endosperm Cotyledon Plumule Radicle
endospermic seed	Main food store is in endosperm. See endosperm.

endospore   A spore formed within a parent cell e.g. bacterial spores. Formed in unfavourable conditions. Cell contents shrink and are enclosed by a thick wall. When suitable conditions return the spore absorbs water, wall bursts and the spores divide by binary fission.

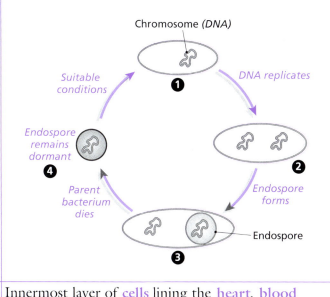

endothelium   Innermost layer of cells lining the heart, blood vessels and lymphatic vessels and makes up the wall of the capillaries. See artery for diagram.

endotherm   Animal whose body temperature is maintained usually at a higher level than that of its surroundings i.e. a warm-blooded animal e.g. human. The source of this heat is from its own metabolism. Compare ectotherm.

energy   (Physics): the ability to do work.

Or

(Biology): contained in chemical bonds of compounds especially adenosine triphosphate (ATP) and carbohydrate and used in metabolism.

energy flow/ transfer	The flow of energy *into* the ecosystem from the sun; and *within* the ecosystem through the different trophic levels along food chains, including detritus, and finally out of the ecosystem into the atmosphere as heat loss due to respiration.
environment	All the conditions in which a cell or organism lives, internal and external, which affect the growth and development of that cell or organism.
enzymes	Proteins with a definitive folded shape. They are highly specific organic catalysts and are affected by heat (temperature), pH, substrate and product concentration. They control the release of energy in respiration and are essential in the energy transfer process of photosynthesis.
enzyme-substrate complex	The combination of the substrate molecule and the enzyme at the active site, prior to their separation and production of the product. 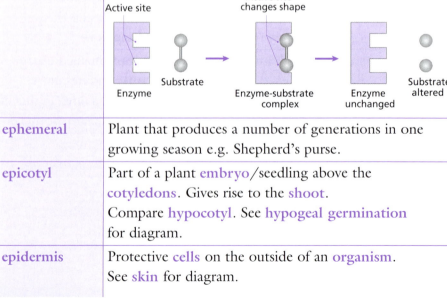
ephemeral	Plant that produces a number of generations in one growing season e.g. Shepherd's purse.
epicotyl	Part of a plant embryo/seedling above the cotyledons. Gives rise to the shoot. Compare hypocotyl. See hypogeal germination for diagram.
epidermis	Protective cells on the outside of an organism. See skin for diagram.

epididymis	Sperm storage organ in the male testes. See the male reproductive system for diagram.
epigeal germination	Cotyledons brought above ground. Hypocotyl elongates e.g. sunflower seed (fruit). Compare hypogeal germination.

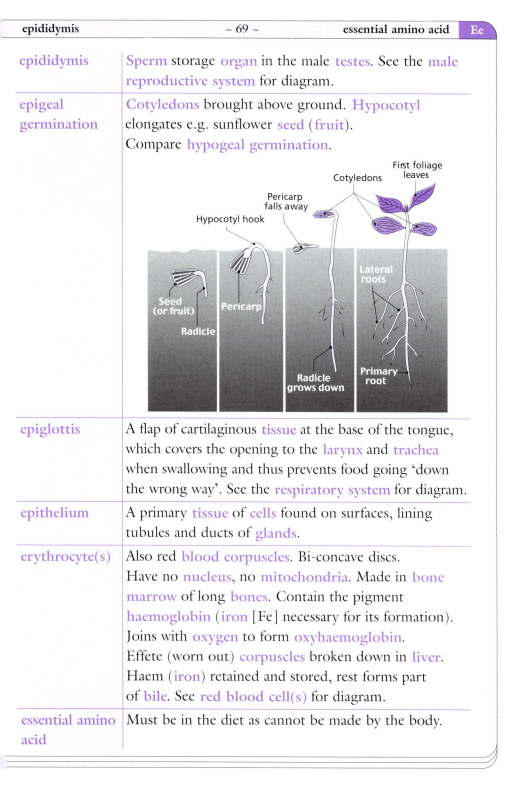

epiglottis	A flap of cartilaginous tissue at the base of the tongue, which covers the opening to the larynx and trachea when swallowing and thus prevents food going 'down the wrong way'. See the respiratory system for diagram.
epithelium	A primary tissue of cells found on surfaces, lining tubules and ducts of glands.
erythrocyte(s)	Also red blood corpuscles. Bi-concave discs. Have no nucleus, no mitochondria. Made in bone marrow of long bones. Contain the pigment haemoglobin (iron [Fe] necessary for its formation). Joins with oxygen to form oxyhaemoglobin. Effete (worn out) corpuscles broken down in liver. Haem (iron) retained and stored, rest forms part of bile. See red blood cell(s) for diagram.
essential amino acid	Must be in the diet as cannot be made by the body.

ethene/ ethylene	Plant growth substance used to promote ripening of some fruits (e.g. bananas, melons, tomatoes), de-greens others (e.g. oranges, lemons and grapefruits) and promotes abscission and inhibits longitudinal growth.
ethyl alcohol	See alcohol.
eukaryote	Cell that has a membrane-bound (true) nucleus. It may also have mitochondria and/or chloroplasts. Compare prokaryote. See cell for diagram.
Eustachian tube	Tube connecting the middle ear to the throat and equalises air pressure on both sides of the eardrum. See the ear for diagram.
eutrophication	A condition where lakes become over-enriched with nutrients, resulting from excess artificial fertilisers washed into rivers and lakes. There is a rapid increase in the growth of alga (algal bloom) as they use up the nutrients. When all the nutrients are used up the algae die and are broken down by bacteria, which use up the oxygen in the water resulting in the death of aquatic organisms such as fish.
evaporation	Loss of water as vapour or the change of state from liquid to gas. Occurs at all temperatures.
evolution	This is about how modern plants and animals have gradually developed from pre-existing forms over long periods of time i.e. changes in species by natural selection in response to environmental changes. Survival of the fittest by adaptation, variation and over-reproduction.

excretion	The elimination of the waste products of metabolism from a cell, tissue or organ:

ORGAN	EXCRETORY PRODUCTS
skin	water, salt, carbon dioxide, urea (sweat)
lungs	carbon dioxide and water vapour
kidneys	water, urea, uric acid, Na, Cl, K
stomata (in plants)	oxygen, water vapour

excretory system	Collection of organs or structures that function in the elimination of the waste products of metabolism, e.g. contractile vacuole (*Amoeba*), kidney, lungs, etc.
exergonic	Chemical reactions that liberate energy, respiration. Compare endergonic.
exhalation	The process of breathing out air from the lungs. See expiration.
exocrine gland	A gland which delivers its secretion through a tube or duct. Does not depend on bloodstream e.g. liver, pancreas, sweat glands and salivary glands. Compare endocrine gland.
exons	Parts of the DNA molecule which code for proteins. Compare interons.
exoskeleton	Found outside body, made of chitin (a strong structural polysaccharide) e.g. on the cockroach. Compare endoskeleton.
expiration	This is a passive process i.e. requires no energy. This is expelling air from the lungs. The diaphragm and intercostal muscles relax causing the ribs and diaphragm to return to their original positions. These actions result in a decreased volume of the thorax (chest cavity) and an increase in the air pressure in the lungs, consequently air is expelled from the lungs. See inspiration for diagram.

exponential phase	See Log phase for diagram.
extensor	A muscle that straightens a joint. See antagonistic muscle pairs for diagram. Compare flexor.
exteroceptors	Sensory receptors that respond to external changes in the environment, e.g. taste receptors. See interoceptors.
extinct	Refers to a species of plant or animal that has died out.
extracellular fluid	All cells in the body are bathed in a fluid called tissue fluid or extracellular fluid (ECF). It is similar to blood plasma but without the plasma proteins. Substances in the blood diffuse from capillaries into the extracellular fluid and then into the cells. Cellular waste products diffuse in the reverse direction. See lymph for diagram.
eye	Organ of vision or sight in animals.

Tear gland — Upper eyelid — Conjunctiva — Aqueous humour — Cornea — Pupil — Lens — Eyelash — Iris — Suspensory ligament — Ciliary muscle — Ciliary body — Tear duct — Sclera — Choroid — Retina — Vitreous humour — Fovea — Blind spot — Optic nerve — External muscle

Ff

F₁ generation	The first filial (daughter) generation. Offspring from crossing the parental generation (homozygous parents differing in one or more characteristics or traits).
F₂ generation	The second filial (daughter) generation. Offspring from crossing members of the F₁ generation with each other.
facultative anaerobe	Organism that respires with or without oxygen, depending on its availability.
faeces	Undigested and indigestible material passed from the body through the anus.
fallopian tube	Oviduct in female mammals. Tube that joins the ovary and the uterus and transfers the egg to the uterus. Fertilisation usually occurs here. See the female reproductive system for diagram.
family planning	Using birth control methods (contraception) to plan the number of children.
fat	Contains the elements carbon, hydrogen and oxygen, but in a different ratio from carbohydrates. A type of lipid that is solid at room temperature.
fatty acid	An acid occurring in or derived from natural fats, waxes, etc. Component of fats or lipids. The smallest lipid is a triglyceride. It is made from three fatty acid molecules, and one glycerol molecule.
fauna	The animals in a locality or region. Compare flora.
female reproductive system	All the organs involved in gamete formation, fertilisation and development of the young in the female.

Fallopian tube
Funnel
Ovary
Ovarian ligament
Uterus
Lining of uterus (endometrium)
Wall of uterus
Cervix
Vagina
Vulva

fermentation

Production of alcohol from starch and sugars (see anaerobic respiration) contained in grain or fruit. This was probably the first use of biotechnology by humans.

fertile period

A short time during which it is possible for a sperm to fertilise an egg e.g in a 28 day menstrual cycle, ovulation occurs on day 14. Sperm can survive for about 48 hours in the female body. The egg is capable of being fertilised for about 48 hours after ovulation. If sexual intercourse occurs between days 12 and 16 it is possible that fertilisation could occur.

fertilisation

The union of a haploid male gamete with a haploid female gamete resulting in the formation of a diploid zygote. In the human female this occurs about halfway along the fallopian tube. See cross-fertilise and self-fertilise (of plants).

Egg

2 Sperm swarm around egg

Path of sperm

1 Insemination occurs here

3 Head of sperm enters egg

5 Fertilisation

4 New membrane forms to prevent further entry of sperm

fibre

Thread-like structure.

Or

Food containing a large content of indigestible material (cellulose) e.g. vegetables. Provides bulk. Eating fibre makes you feel full so can prevent you overeating as part of a balanced diet. Gives the muscles of the gut wall something to push against. Keeps the contents of the gut moving. Absorbs water, keeps faeces soft, easier to egest. Prevents constipation. Helps prevent bowel (colon) cancer caused by carcinogens produced by bacteria in the colon. Carcinogens are then present in faeces – fibre increases the bulk of the faeces and dilutes the carcinogens.

fibroids	Benign (non-cancerous) growths in the endometrium. They range in size from a pea to an orange. Consist of muscle and connective (fibrous) tissue and grow slowly in the uterine wall. Mostly occurring in women over 30, often multiple and cause discomfort. Thought to be associated with the levels of oestrogen. Oral contraceptives containing oestrogen can cause fibroids to enlarge. Large fibroids may cause uterine lining to wear away resulting in heavy menstrual bleeding and loss of iron which leads to anaemia. Large fibroids causing complications can be surgically removed.
fibrous	Made of or containing thread-like structures.
fibrous root	Group of thread-like roots approximately equal in size arising from base of stem e.g. grasses. Fibrous roots
filament	A thread-like structure or form of growth. Or The stalk of the anther (pollen producing part of flower). See flower for diagram.
fixed joint	Bones of joint fused together e.g. sutures of skull, coccyx. Allows no movement.

flaccid	Lacking rigidity. A cell that is flaccid is capable of taking in more water by osmosis e.g. cells in a wilting leaf. Achieved by placing a plant cell in a solution that is hypertonic to the cell sap. Water flows out of the vacuole and the cell membrane shrinks. See salting and sugaring – methods of food preservation.
flagellum	(Plural = flagella.) Whip-like process or appendage of a cell e.g. bacteria. Used for propulsion. See bacterium for diagram.
flaming	Procedure used when growing micro-organisms. Open ends of containers and inoculating loops are passed through a flame to kill any foreign, unwanted micro-organisms and to prevent contamination of the culture.
flexor	A muscle that bends a joint. Compare extensor. See antagonistic muscle pairs for diagram.
flora	The plants of a locality or region. Compare fauna.
flower	Reproductive shoot of plant, containing sepals, petals, stamens and carpel.

foetus	Developing embryo with the appearance of the fully developed animal. In humans from eight weeks after fertilisation until birth.

folic acid	(Vitamin B complex.) Water soluble, essential for the formation of RNA and DNA and the normal production of red blood cells. Deficiency causes anaemia during pregnancy and certain birth defects e.g. spina bifida. Source: liver, green vegetables and yeast.
follicle	Fluid-filled sac in the ovary in which the egg develops.
follicle stimulating hormone	See FSH.
food chain	A list of species such that each is food for the next species in the list. A chain of organisms through which energy is transferred. Begins with green plants (producers) eaten by small animal (primary consumer), then a larger animal (secondary consumer), etc. The final member of the chain is the tertiary consumer.
food preservation	Method used to prevent food from decomposition, fermentation or deterioration e.g. salting and sugaring.
food pyramid	Pyramid of five levels. Each level represents a different type of food. From the bottom up these are: cereals and starches (6+); fruit and vegetables (4+); dairy products (3); meat, fish, eggs, beans (2); sweets, chocolates, cakes (sparingly). The numbers in brackets represents the suggested minimum number of daily servings.

food web	A chart showing all the feeding connections in the habitat. Constructed by showing the links between all the interconnecting food chains in the habitat. 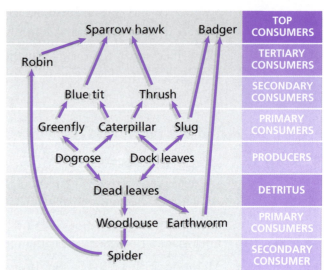
force-filter	See pressure filtration.
fossil	The remains of organisms that lived long ago. An isolated fossil has no value, but examination of fossils found in different layers of sedimentary rocks shows that: • lower layers of rock contain different fossils from the layers above • the more recent fossils (found nearer surface) are more complex than the earlier ones • many of the early fossilised plants and animals do not exist today, showing extinction • many of the plants and animals alive today have not been found as fossils, showing new variations/species.
fossil fuel	Coal, oil, natural gas and turf formed from the remains of dead plants and animals over long periods of time.

fovea	Area of most acute vision on retina of eye, directly behind centre of lens. Has a high density of cones. See the eye for diagram.
fruit	A mature ripened ovary or modified part of a flower (e.g. the receptacle), usually containing seeds.
FSH	**Follicle stimulating hormone**. Sex hormone produced by the anterior lobe of the pituitary. Stimulates sperm production in the male and egg and oestrogen production in the female.
fungus	(Plural = fungi.) Plant that has no stamens or pistils (hence no proper flowers or seeds) and no chlorophyll. **Eukaryotic** in nature. Feeds on organic matter e.g. mushrooms, yeasts, toadstools and moulds. Found in a wide variety of habitats e.g. salt water, fresh water, terrestrial, airborne.
fused joint	Bones of joint fused together e.g. sutures of skull, coccyx. Allows no movement.

Gg

gall bladder	A sac-like vessel that lies behind the lobes of the liver and stores bile, which is produced by the liver. The bile duct drains the gall bladder. See the bile duct for diagram.
gamete	Any cell which must fuse with another cell in order to produce a new individual. Haploid reproductive cell i.e. contains half the required genetic information for the formation of a new individual or half the somatic number of chromosomes (in humans = 23) e.g. sperm cells, egg.
ganglion	(Plural = ganglia.) A mass of nervous tissue, containing the cell bodies of the neurons, lying outside the central nervous system (CNS i.e. brain and spinal cord). Or A collection of cell bodies within the peripheral nervous system (PNS) e.g. the dorsal root ganglion.

Ganglion
Myelin sheath
Nerve fibre
Axon
Fat
Bundles of neurons
Neurotransmitter swelling

gaseous exchange	At respiratory surfaces (lungs in humans, gills in fish) it is the removal of waste gas (carbon dioxide) and the taking in of necessary gas (oxygen). The gaseous exchange is the reverse through the stomata of the leaves of plants during photosynthesis. See alveolus for diagram.

gastric juice	Collective term for all secretions by the glands in the stomach e.g. mucus, enzymes and hydrochloric acid.
gene	Unit of heredity found on a chromosome, and is an instruction to the cell to make a particular substance, which helps regulate a trait (characteristic) of an organism. See locus for diagram.
gene expression	The process of changing the information in a gene into a protein and the effect that protein has on the organism.
gene mutation	Change in a single gene i.e. a change in the DNA structure. Changes the nucleotide sequence, changes the amino acid sequence, which in turn changes the protein produced e.g. sickle cell anaemia, phenylketonurea (PKU).
gene pool	All the genetic information possessed by members of a population.
gene therapy	The insertion of genetically altered or normal genes into cells using recombinant DNA technology, to replace the defective genes that cause genetic disorders.
general defence system	Prevents the entry of microbes (by acting as a barrier) and destroys microbes once they get inside the body. Composed of the skin, blood clotting, respiratory system, digestive system, body fluids, and phagocytes.
generative nucleus	Found in the pollen grain and divides by mitosis after pollination to form two male gamete nuclei.
genetic code	Controls the metabolic reactions of cells. Contained on the chromosomes in the sequence of nitrogenous bases. See triplet for diagram and protein synthesis for more information.
genetic disorder	Condition resulting from a change in a gene or the number of chromosomes. See mutation.

genetic engineering	Modern techniques or processes used to artificially alter the genetic information in the chromosome of an organism. The process involves the following: (a) isolation of gene (b) cutting (restriction) (c) transformation (ligation), introduction of base sequence changes (d) expression. Applications of genetic engineering include: • tomato plants: gene for producing the enzyme needed to soften the fruit on ripening has been altered and no longer functions. Fruit remains hard, easier to harvest, used to make tomato ketchup. • sheep: have been given the human gene for factor VIII. Factor VIII is a substance needed for blood clotting. Persons with haemophilia are missing this gene. It is hoped that the factor viii will be able to be extracted from the sheep's milk. • micro-organisms: the production of human insulin by bacteria.
genetics	Study of the mechanism of inheritance and variation of traits or characteristics as transmitted from one generation of plants or animals to another. Or the study of the structure and function of genes and their transmission from parents to offspring.
genetic screening	Tests to identify the presence or absence of changed or harmful genes possessed by an individual.
genome	All the genes of an individual or all the genes possessed by members of a population or species.
genotype	Genetic make-up of an individual or the genes that they inherit e.g. Tt.
genus	(Plural = genera.) A grouping in the classification of an organism between family and species.

geotropism	The growth response of a plant to gravity.
germ layer	A portion of a group of cells capable of developing into the various different tissues and organs of a new individual. Cells in the human blastocyst arrange themselves into three layers and give rise to the following tissues:

- ectoderm – skin, hair, nails
- mesoderm – muscles, skeleton, kidneys
- endoderm – linings of the alimentary canal, trachea and bronchi.

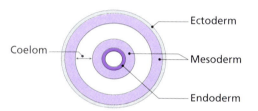

germination	Is the beginning of growth of seeds, spores or pollen grains after a period of dormancy. Certain conditions must be available i.e. water, oxygen and a suitable temperature.
gestation period	Length of time from conception to birth i.e. length of pregnancy, carrying foetus in womb (uterus). In humans = 40 weeks or 266 days.

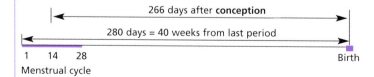

gibberellin	Plant growth promoter which stimulates cell elongation and germination.

gill(s)	Respiratory organ/surface in some aquatic organisms through which gaseous exchange takes place. Or The reproductive structures radiating from the centre on the underside of mushrooms.
gland	Organ in the body of an animal that secretes substances either for use by the body or for elimination (excretion).
gliding joint	A bone joint found, for example, at the wrist and ankle: allows limited circular movement.
glomerular filtrate	Liquid containing dissolved substances that pass from the glomerulus, in the nephron of the kidney, into Bowman's capsule. Similar to blood plasma but without the proteins (albumen, fibrinogen, prothrombin). See Bowman's capsule for diagram.
glomerulus	Small bunch of capillaries projecting into Bowman's capsule in the nephron of kidney. Pressure filtration occurs here. See Bowman's capsule for diagram.
glottis	The opening of the larynx. See respiratory system for diagram.
glucagon	A hormone released by the pancreas. In humans glycogen is broken down to glucose through the action of the hormones (adrenaline and glucagon), as more glucose is demanded for respiration. Glucagon and insulin work together to control the level of sugar in the blood.
glucose	$C_6H_{12}O_6$. A 6-carbon single sugar, monosaccharide, hexose sugar. Formed in plants during photosynthesis and is the end product of the digestion of carbohydrates. Most living organisms need it for respiration.
glycerol	Fats are formed from one glycerol molecule attached to three fatty acids. See triglyceride for diagram.

glycogen	Animal storage polysaccharide found in muscles, liver and brain.
glycolysis	The first stage of respiration that involves the conversion of glucose (6-carbon compound) to pyruvic acid (3-carbon compound) in the cytoplasm with the release of energy. Energy is also required to start this process. See pyruvic acid for more information.
goitre	Unsightly enlargement of the thyroid gland due to excessive production of the hormone thyroxine. See thyroxine and Grave's disease (on CD).
gonad	An animal organ that produces gametes e.g. testis in the male and ovary in the female.
Graafian follicle	Fluid filled vesicle in ovary of female mammal containing an egg.
grafting	Method of plant propagation where scion is attached to stock e.g. apple trees.

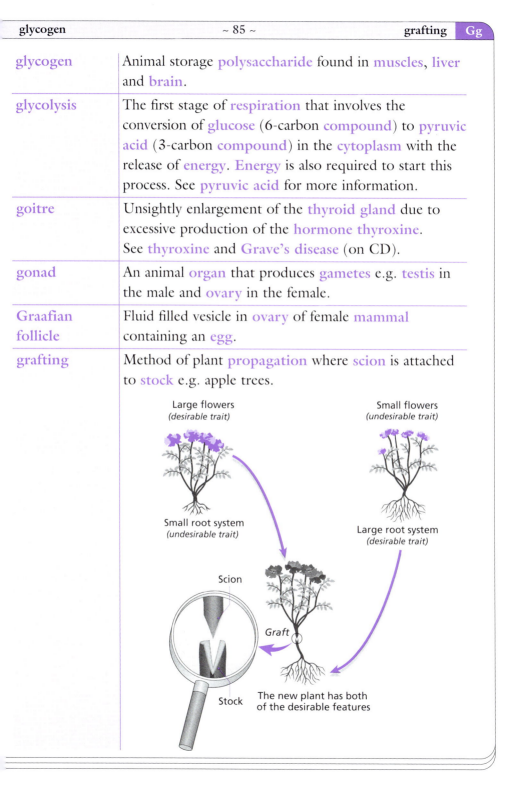

Large flowers
(desirable trait)

Small flowers
(undesirable trait)

Small root system
(undesirable trait)

Large root system
(desirable trait)

Scion

Graft

Stock

The new plant has both of the desirable features

granum	(Plural = grana.) Found in chloroplasts stacked in piles (like coins). It is here that the light stage/phase of photosynthesis takes place. See chloroplast for diagram.
grazing food chain	A food chain where the initial plant is living e.g.: (i) grass → grasshoppers → frogs → hawks (ii) honeysuckle → aphids → ladybirds → thrushes (iii) seaweed → winkles → crabs → herring gulls (iv) phytoplankton → zooplankton → copepod → herring.
greenhouse effect	Gases in the atmosphere, such as carbon dioxide, trap heat escaping from the earth and radiate it back to the surface. The gases are transparent to sunlight but not to heat and act like the glass in a greenhouse.
grey matter	Found in the spinal cord and brain. Consists of nerve cell bodies and dendrites.
ground tissue	Living plant cells making the soft parts of leaves e.g. pith, cortex and spongy tissue of leaves. Living cells, thin walled and contain a vacuole, may contain chloroplasts for photosynthesis or other plastids for food storage.

growth	Increase in the size of an organism due to cell division. Or Increase in the size of a population due to the birth rate greater than death rate.
growth curve	Graph showing population size of micro-organisms over a period of time. Has five phases: lag phase, log phase, stationary phase, death phase (decline phase) and survival phase.
growth hormone	Produced by anterior (front) lobe of pituitary. Stimulates body growth – an excess results in giantism and a deficiency results in dwarfism. Compare plant growth regulators.
guanine	$C_5H_5N_5O$ – one of the purine nitrogenous bases present in DNA and RNA.
guard cell	Found each side of stomata in the epidermis of plants. Controls opening and closing of stomata and contains chlorophyll. Regulates gaseous exchange and transpiration. See stoma for diagram.
gymnosperm(s)	Vascular plants which produce naked seeds e.g. conifers (produce cones). Evergreen.

Hh

habitat	The particular place within the ecosystem where an organism lives and to which it is adapted.
haemoglobin	Oxygen-carrying respiratory pigment containing iron. Present in red blood corpuscles of animals. Joins with oxygen to form oxyhaemoglobin.
haemolysis	The bursting of red blood corpuscles with the release of haemoglobin. Achieved by placing red blood corpuscles in a solution, which is hypotonic to the cell sap. Compare crenation.
haemophilia	Hereditary sex-linked (see sex linkage) disease of males in which blood clotting is defective and prolonged bleeding occurs. Gene responsible transmitted from mother (carrier) to son.
haploid (n)	Having one of each of the pairs of chromosomes characteristic for a species Or half the somatic number of chromosomes Or a single set of unpaired chromosomes. Compare diploid. Chromosomes n = 2 n = 3 n = 4
hay fever	A seasonal allergy, symptoms include sneezing fits, a blocked or runny nose, a tickle in the roof of the mouth and itchy watery eyes. Caused by the pollen of certain wind-pollinated plants e.g. grass.
HCG	Hormone secreted by an implanting fertilised egg (blastocyst). HCG maintains the corpus luteum in the ovary (until the ninth week of the gestation period, after which the placenta takes over) and prevents the breakdown of the endometrium. The analysis of urine for the presence of HCG is used to confirm pregnancy.

heart	A hollow muscular organ that contracts rhythmically to pump blood around the body.

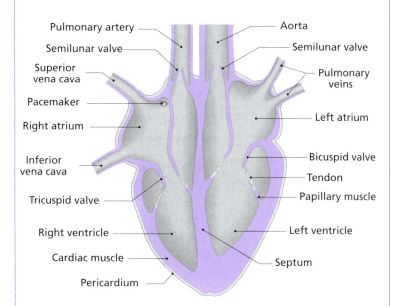

Pulmonary artery — Aorta
Semilunar valve — — Semilunar valve
Superior vena cava — — Pulmonary veins
Pacemaker —
Right atrium — — Left atrium
Inferior vena cava — — Bicuspid valve
— Tendon
Tricuspid valve — — Papillary muscle
Right ventricle — — Left ventricle
Cardiac muscle — — Septum
Pericardium —

heart attack	Medical term = myocardial infarction. May be caused by a build-up of cholesterol in the cardiac arteries. This stops oxygen getting to the heart tissue and results in the death of some of the heart muscle. Symptoms include severe chest pain, sweating and shortness of breath. If the heart attack causes a complete stoppage of the heart, this is a cardiac arrest and may cause the person to die.
heartbeat	Pulsation of heart, controlled by pacemaker (found in wall of right atrium).
hedgerow	Row of wild bushes and plants forming a hedge and is the habitat of a variety of birds, insects and small animals.
helper T-cells	A type of lymphocyte that stimulates the production of antibodies by causing the B-cells to reproduce when an antigen is present. Also stimulates other T-lymphocytes (killer and suppressor) to work and enhance their effectiveness.

hepatic	Of or pertaining or relating to the liver.
hepatic artery	Blood vessel carrying blood from the heart to the liver. See hepatic portal vein for diagram.
hepatic portal vein	Large vein that drains the digestive system and carries absorbed food (glucose, amino acids, minerals and vitamins) from the villi to the liver.

Hepatic vein

Unwanted amino acids

UREA

STORED
Glycogen
Minerals
Vitamins

Hepatic artery

Liver

Ileum

Hepatic portal vein—
(rich in absorbed food)

hepatic vein	Blood vessel carrying blood from the liver towards the heart. See above for diagram.
hepatitis	Inflammation of the liver usually caused by a virus, alcohol poisoning or an adverse drug reaction. Symptoms include jaundice and fever.
herbaceous	Refers to a plant that does not develop persistent woody tissues (xylem) i.e. has a non-woody stem. Aerial parts of herbaceous plants usually die back in winter.
herbivore	An animal that feeds exclusively or mainly on plants. A plant eater e.g. rabbit. Compare carnivore, omnivore.
heredity	The natural law or property of organisms whereby their offspring have various physical and mental traits of their parents or ancestors i.e. certain traits, controlled by a genetic code within the chromosomes, are transmitted from one generation to the next.

heterotroph	An organism that cannot make its own food. Depends on other organisms as sources of food e.g. all animals, saprophytes and parasites.
heterozygote	An individual that does not breed true for a particular trait, since it has a pair of dissimilar genes for the trait e.g. Nn, Tt. Compare homozygous.
hinge joint	Type of synovial joint e.g. elbow, knee. Allows movement in one plane only. See joint(s) for diagram.
HIV	Human immunodeficiency virus – the virus that attacks the T lymphocytes in the human immune system and destroys the body's defences. Causes the development of AIDS (Acquired Immune Deficiency Syndrome).
homeostasis	The maintaining of a constant internal environment (i.e. concentrations of water, salt, turgidity, temperature, etc.) of a cell or organism or the processes involved with this. Usually achieved by diffusion and the respiratory and excretory systems.
homologous chromosomes	Pair of chromosomes in diploid cell, same length and shape and contain genes at the same position (locus) for the same traits. See diploid for diagram.
homologous structure	Same structure found on different organisms, originating from a common ancestor, but with different functions e.g. pentadactyl limb (forelimb). See comparative anatomy.
homozygous	An organism that breeds true for a particular trait, because it possesses a pair of similar genes for the trait e.g. TT or tt. Compare heterozygote.
hormone	A substance (chemical message) produced by an endocrine gland secreted directly into and transported by the bloodstream to other parts (target organs) of the body where it evokes a response. Hormones regulate metabolic activity. A hormonal response generally works slowly over a longer period of time compared with a nervous response.

host

An animal (organism) in or on which another organism lives and derives nourishment e.g. cow or sheep is host to the liverfluke.

Or

Animal that has received a transplanted organ.

Or

Cell in which a virus replicates or into which something has been inserted.

host resistance

The capacity of an organism (the host) to withstand the action or effects of a pathogen or other chemicals or disease.

human skeleton

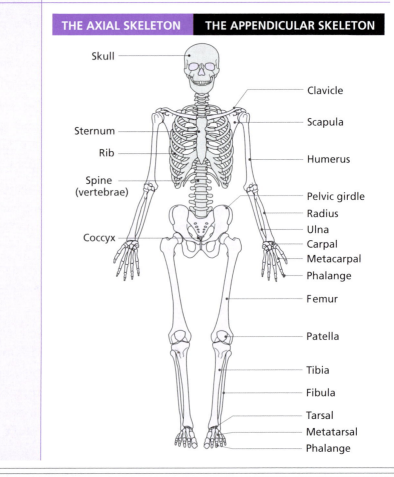

THE AXIAL SKELETON   THE APPENDICULAR SKELETON

Skull

Clavicle

Scapula

Sternum

Rib

Humerus

Spine (vertebrae)

Pelvic girdle

Radius

Ulna

Coccyx

Carpal

Metacarpal

Phalange

Femur

Patella

Tibia

Fibula

Tarsal

Metatarsal

Phalange

humerus	The bone between the shoulder and elbow in humans. See arm for diagram.
humidity	The amount of water vapour in the air, usually expressed as a percentage.
humus	Dead and decaying plant or animal material, organic matter.
hyaline cartilage	Clear cartilage found in septum of nose, trachea and covering moving ends of bones. See synovial joint for diagram.
hybrid	A heterozygous individual. Or A crossbred animal or plant i.e. one arising from a cross between two different species, races or varieties.
hydrogen bonds	Weak chemical bonds/attachments formed between parts of molecules that have a slight charge. Found between the nitrogenous base pairs (adenine and thymine, and guanine and cytosine) and are responsible for holding the two strands of the DNA molecule together. Compare covalent bond and ionic bond.
hydrogen carrier(s)	Substances that transport hydrogen in biological reactions by reduction and oxidation e.g. (a) during the light stage of photosynthesis, NADP is reduced to $NADPH_2$ and is carried into the dark stage where it is oxidised and the NADP returns to the light phase to be reduced again, and (b) during oxidative phosphorylation of respiration, hydrogen is passed between four carriers, the final one giving it to oxygen to form water.
hydrolysis	Splitting, disintegration or decomposition by chemical reaction with water e.g. $$C_{12}H_{22}O_{11} \ + \ H_2O \ \rightarrow \ C_6H_{12}O_6 \ + \ C_6H_{12}O_6$$ maltose + water → glucose + fructose
hydrophilic	Water loving. Attracted to water or capable of being wet by water. Compare hydrophobic.

hydrophobic	Water hating. Repels water or has an aversion to water. Compare hydrophilic.
hydrotropism	The growth response of a plant to water.
hypermetropia	See hyperopia.
hyperopia	Long sightedness. Cannot see objects close to the eye, eyeball too short. Corrected with convex lenses. Compare myopia.
hypertonic	A solution with a higher solute concentration than the surrounding solution. Compare hypotonic and isotonic.
hypha(e)	Filament in mycelium of fungus. See *Rhizopus* for diagram.
hypocotyl	Part of a plant embryo/seedling below the cotyledons. Gives rise to the radicle, produces the root. Compare epicotyl.
hypogeal germination	Cotyledons remain below ground, epicotyl elongates e.g. broad bean. Compare epigeal germination.
hypothalamus	Part of the brain (above the pituitary gland) responsible for appetite, sleep, osmoregulation, body temperature. Maintains homeostasis. See the brain for diagram.
hypothesis	Groundless assumption taken from known facts (educated guess) as a starting point for further investigations.
hypotonic	A solution with a lower solute concentration than the surrounding solution. Compare hypertonic and isotonic.

Ii

IAA	Indoleacetic acid. A plant growth hormone (auxin) that stimulates cell elongation.
immobilised enzyme	Enzyme trapped in beads or gel so that it will react with, but not mix with, its substrate. This makes it easy to recover the enzyme from the product and it is reusable.
immovable joint	A joint whose bones interlock and prevent it from moving e.g. bones of the cranium.
immune response	The production by the body of a reaction to infection. The primary immune response takes place when the foreign protein (antigen) and antibody meet. A secondary immune response takes place on subsequent encounters with the same antigen.
immune system	Protective methods the body has to keep it safe from disease or disease causing organisms (pathogens) and provides defence against the growth of cancer cells. Composed of the spleen, thymus gland and lymph nodes.
immunisation	Rendering immune. See inoculation.
immunity	The ability of the body to resist infection.
impermeable	Cannot be penetrated by solids or liquids particles.
implantation	Process by which the blastocyst attaches itself to and becomes embedded in the lining of the uterus (endometrium). This process may be attempted artificially as part of in vitro fertilisation for the treatment of infertility.
impulse	Chemical change transmitted along a nerve fibre, which brings about a response in an effector (muscle or gland).
incomplete dominance	In the heterozygous condition both alleles show complete dominance and an intermediate phenotype results e.g. in shorthorn cattle red x white, F_1 = roan. Also called co-dominance.

incubate	Cause to develop/grow by supplying all the necessary requirements e.g. temperature (use oven or incubator), food (as a nutrient medium in the case of bacteria), etc.
incubation period	Length of time necessary to develop. Or Length of time between infection by pathogen and development of first symptoms of illness.
induced fit theory	The active site of an enzyme is not a fixed shape. It can change slightly to better fit the shape of the substrate when forming the enzyme-substrate complex. See enzyme-substrate complex for diagram.
induced immunity	To give the body the ability to fight infection by (a) suffering the illness or (b) by vaccination. Induced immunity can be active or passive.
infection	Disease caught or transmitted by water, air, etc.
infertility	The inability to conceive or produce children. May be caused by a low sperm count or low sperm mobility in males, blocked fallopian tubes in females or endocrine gland failure in either or both. See in vitro fertilisation.
inflorescence	The arrangement of flowers on the stem of a plant – single or group, along the stem or clustered, e.g. lupin, foxglove, etc.
ingestion	The process of taking food into the body.
inhalation	The process of breathing in. See inspiration.
inheritance	The qualities, characteristics, or traits an organism possesses as a result of receiving them from its parents or ancestors.
inoculation	Injection containing a vaccine or weakened form of a disease to stimulate the body to protect itself against the disease. See active immunity.
inorganic	Any complex compounds that do not contain the elements carbon and hydrogen i.e. not organic. Not from living things.

inorganic molecule	Of mineral origin or not organic. Molecules that do not contain carbon-hydrogen bonds e.g. CO_2, H_2O, NaCl.
inspiration	Is an active process i.e. requires energy. This is taking air into the lungs. The intercostal muscles between the ribs contract causing the ribs to move upwards and outwards. The circular muscles of the diaphragm contract and cause it to flatten. These actions result in an increased volume of the thorax and a decrease in the air pressure in the lungs – consequently air is drawn into the lungs.
insulin	A hormone produced by the islets of Langerhans in the pancreas. It controls the blood sugar level, converts glucose to glycogen. A deficiency causes diabetes mellitus.
integuments	Inner and outer layers which surround a mature ovule in a flowering plant to form the testa (seed coat). Small opening remains = micropyle. See ovule for diagram.
inter-	Between or among e.g. intercellular = between or among cells, inter-relationships = relationships between organisms.
intercostal muscle(s)	The muscles located between the ribs. See inspiration and expiration. See the respiratory system for diagram.
interdependence	In a habitat or ecosystem, the reliance of organisms of different species on each other for an essential resource and the ways in which they depend on each other for survival.
interferon	A protein substance produced by cells which have been infected by a virus. It has the effect of stopping virus growth and protecting surrounding cells against viral infection. Can be produced by genetic engineering and is used to treat a variety of conditions (hepatitis B and C) and some cancers.

interneurons	Neurons that carry messages from one or more sensory neurons to motor neurons. They are found within the central nervous system.
internode	The part of a stem between two nodes.
interoceptors	Sensory receptors that respond to changes in the internal environment e.g. stretch receptors in the muscles. See exteroceptors.
interons	Also known as junk genes. Pieces of DNA that do not code for a protein and seem to have no function other than to separate genes. Compare exons.
interphase	A preparatory stage before, or a rooting stage after mitosis or meiosis. Interphase is followed by prophase. See cell cycle for diagram.
interspecific competition	The struggle between organisms of different species for a limited resource. Compare intraspecific competition.
intestine	That part of the alimentary canal after the stomach to the anus i.e. duodenum, jejunum, caecum, colon, rectum and anus. See the digestive system for diagram.
intra-	Inside or within e.g. intracellular = within cells.
intraspecific competition	The struggle between organisms of the same species for a limited resource. Compare interspecific competition.
intrauterine device	A plastic and copper device used as a method of contraception. It is inserted into the uterus (womb) and it prevents implantation of a fertilized egg. It is about 90% effective.
invertebrate	Animal without a backbone or spinal column, e.g. earthworm, jelly-fish.
in vitro fertilisation	A method of fertilisation outside the woman's body (in a 'test tube' or other laboratory environment) used to treat infertility and help the woman conceive.
involuntary	Not under conscious control.

iodine	Non-metallic element needed for the production of thyroxine (hormone). Also used to detect the presence of starch – turns starch a blue/black colour. Also used in medicine as an antiseptic.
ionic bond	A chemical bond between ions of opposite charge (e.g. Na^+ + Cl^- → NaCl). Electrostatic attraction holds the ions together. Compare covalent bond and hydrogen bond.
iris	Coloured part of the eye in front of the lens. Controls the amount of light entering the eye. See the eye for diagram.
iron	Metallic element, symbol = Fe, needed for the production of haemoglobin and some enzymes. A deficiency causes anaemia.
islets of Langerhans	A group of cells in the pancreas that produce the hormone insulin.
isotonic	A solution with the same solute concentration as another solution. Compare hypertonic and hypotonic.
IUD	See intrauterine device.
IVF	See in vitro fertilisation.

Jj

joint(s)   The place where two **bones** meet. Types of joints include **immovable joints**, **slightly movable joints**, freely moving and **synovial joint**s.

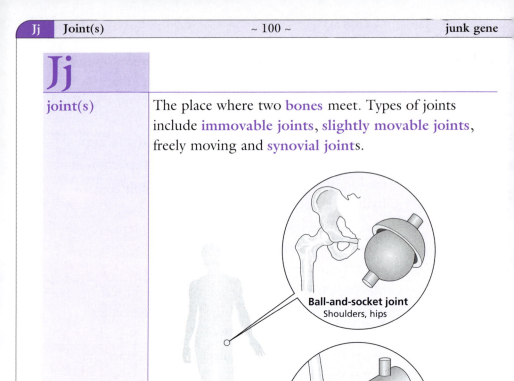

Ball-and-socket joint
Shoulders, hips

Hinge joint
Elbows, fingers, knees

joule   Unit of work and **energy**. One joule = work done when a weight of one newton moves a distance of one metre. One joule of energy is required to do one joule of work.

junk gene   Portions of **DNA** that do not code for **protein** and seem to have no function other than to separate **genes**. See **interons**. Compare **exons**.

Kk

karyogram	Diagram of chromosomes in a cell arranged in pairs and in order of size.
karyotype	Character/make-up of cell nucleus determined by the type of chromosomes present.
keratin	Tough fibrous protein with a high sulphur content forming structures such as horns, nails, hair, feathers etc.
kidney	One of a pair of excretory organs that form and extract urine. Also involved in osmoregulation.

kidney stones	Caused by concentrated substances in the urine slowly crystallising to form stones in the urine-collecting part of the kidney. May also form in the ureters or bladder. Small stones may be excreted with the urine, larger ones may damage ureters and block the flow of urine.
killer T-cells	Lymphocytes that attack large pathogens e.g. unicellular parasites. They destroy cancer cells and cells containing viruses. Produce a chemical called perforin that punctures cell membranes.

kilojoule	1,000 joules.
kingdom	One of the five main groupings into which all living things are classified: animal kingdom, plant kingdom, fungus kingdom, protist kingdom (see protista) and monera (prokaryote) kingdom.
Kreb's cycle	Citric acid cycle. The second stage in respiration involving a complex series of enzyme-controlled reactions where pyruvic acid, in the presence of oxygen, is broken down to carbon dioxide, water and ATP. This takes place in the lumen of the mitochondria.

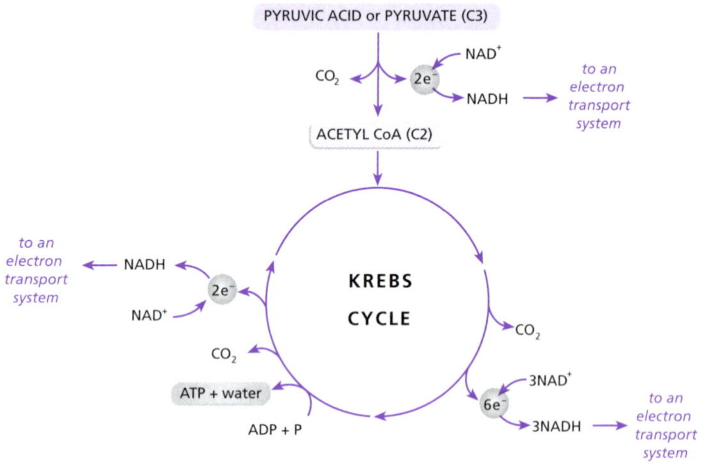

PYRUVIC ACID or PYRUVATE (C3)

NAD⁺

CO_2 2e⁻   to an electron transport system

NADH

ACETYL CoA (C2)

to an electron transport system ← NADH ←

2e⁻

NAD⁺

CO_2

KREBS CYCLE

CO_2

ATP + water

ADP + P

3NAD⁺

6e⁻   to an electron transport system

3NADH

Ll

labour	Pains of giving birth (**parturition**) caused by contractions of the **uterus**.
lactation	**Milk** production by the breasts of **mammals**.
lacteal	A **lymphatic vessel** in a villus of the **small intestine**. Absorbs **fatty acids** and **glycerol** after **digestion** of fats. See **villis** for diagram.
lactic acid	$CH_3CH(OH)COOH$. A by-product of **anaerobic respiration** in animal **cells**. Can be converted back into **pyruvic acid** and then oxidised in the usual way if **oxygen** is present, or it may be sent to the **liver** where, in the presence of **oxygen**, it is converted to **glycogen** or **glucose**.
lactose intolerance	An inherited genetic condition where individuals affected do not have the **enzyme** lactase, which breaks down lactose to **glucose** and galactose.
lag phase	The first phase of a **growth curve**. The initial settling-in stage when an **organism** is introduced into a new **environment**, during which no **growth** occurs.
lamina	**Leaf** blade.

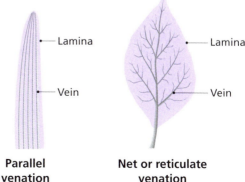

Parallel venation   Net or reticulate venation

large intestine — That part of the **alimentary canal** composed of the **caecum**, **colon**, rectum and anus. It functions in the **absorption** of **water** and **mineral salts** back into the **blood**, which prevents the body becoming dehydrated. Waste (**faeces**) becomes more semi-solid, passes into rectum and is stored, eventually released through anus. This is **egestion**. See the **digestive system** for diagram.

larynx — Voice box, located at top of **trachea**. See the **respiratory system** for diagram.

lateral — Of, from, towards or at the side of something.

lateral bud — **Bud** found on the side of a winter twig or **bulb**. Compare **axillary bud**. See **leaf** for diagram.

law of independent assortment — States that during **gamete** formation each member of a pair of **genes** may combine randomly with either of another pair.

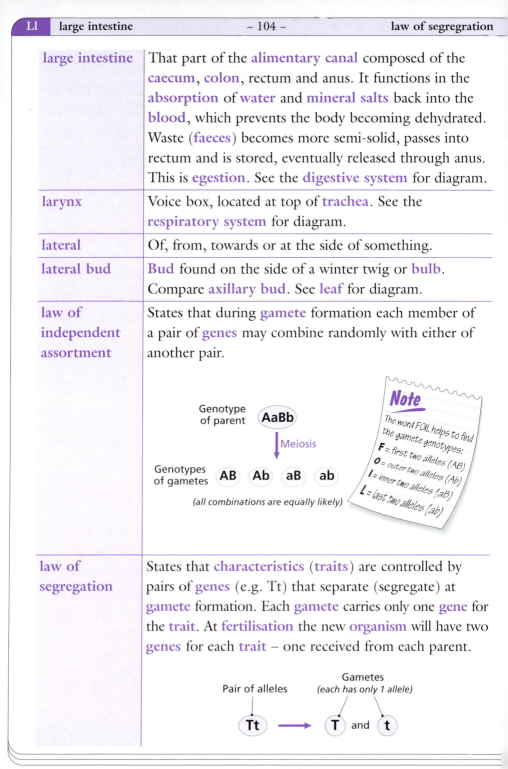

Genotype of parent **AaBb**

↓ Meiosis

Genotypes of gametes **AB** **Ab** **aB** **ab**

(all combinations are equally likely)

Note

The word FOIL helps to find the gamete genotypes:
F = first two alleles (AB)
O = outer two alleles (Ab)
I = inner two alleles (aB)
L = last two alleles (ab)

law of segregation — States that **characteristics** (**traits**) are controlled by pairs of **genes** (e.g. Tt) that separate (segregate) at **gamete** formation. Each **gamete** carries only one **gene** for the **trait**. At **fertilisation** the new **organism** will have two **genes** for each **trait** – one received from each parent.

Pair of alleles

Gametes *(each has only 1 allele)*

Tt ⟶ **T** and **t**

leaf	Outgrowth from stem which functions in photosynthesis, transpiration and respiration. Size can vary from large and flat to needle-like.

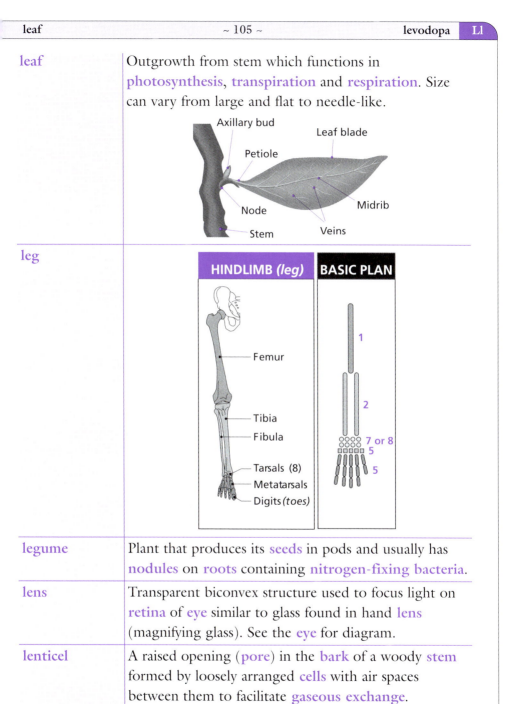

legume	Plant that produces its seeds in pods and usually has nodules on roots containing nitrogen-fixing bacteria.
lens	Transparent biconvex structure used to focus light on retina of eye similar to glass found in hand lens (magnifying glass). See the eye for diagram.
lenticel	A raised opening (pore) in the bark of a woody stem formed by loosely arranged cells with air spaces between them to facilitate gaseous exchange.
levodopa	Drug used in the treatment of Parkinson's disease. It is converted into dopamine by the body.

LH	**Sex hormone** produced by the posterior lobe of the **pituitary**. Causes **ovulation** in the female on day 14 of a 28-day **menstrual cycle** and stimulates the **corpus luteum** to produce **progesterone**. In the male it stimulates the **testes** to produce **testosterone**.
lichen	A symbiotic association of a **fungus** and an **alga**.
life	Something that all living things have. Difficult to quantify. Involves the interaction of the **characteristics of life**.
ligament	An elastic connective **tissue** that joins **bone** to **bone**. See **synovial joint** for diagram.
light stage (of photosynthesis)	Takes place in the **grana** of the **chloroplasts**. Its functions are: • to produce **ATP** • to produce hydrogen atoms. Both of which will be used in the **dark stage** for the reduction of carbon dioxide and production of **carbohydrate**. See **photosynthesis** for diagram.
lignin	Strengthening material found in **cell walls** of woody **tissue**. See **xylem** for diagram.
linkage	**Genes** on the same **chromosome** that are not separated at **gamete** formation and are inherited together. Complete linkage, i.e. **genes** never separating, seldom occurs. The closer the **genes** are on a **chromosome** the greater the degree of linkage; the further apart, the lesser the degree of linkage. In the diagram R and S are linked; as are r and s.

lipase	An enzyme produced in the stomach (gastric lipase), pancreas (pancreatic lipase) and small intestine (intestinal lipase), which converts lipids to fatty acids and glycerol during the process of digestion. Its optimum pH is slightly basic in the pancreas and small intestine but acidic in the stomach.
lipid	Fat or fat-like substance. Contains the elements carbon, hydrogen and oxygen, but not in the same proportion as sugars. Examples of fat are oils, waxes and fat. Composed of fatty acids and glycerol. Smallest lipid is a triglyceride. Lipids are important in the following ways:

(i)   storage molecules
(ii) insulation: under the skin
(iii) protective properties: around body organs
       e.g. kidney
(iv) structural component: as phospholipids in
       cell membranes.

liver	Large exocrine gland and organ on right hand side of body below the diaphragm and beside/over the stomach. Has a range of functions:

- produces bile
- breaks down old red blood cells (but keeps and stores the iron for making new red blood cells)
- detoxifies the blood (drugs and poisonous substances are made harmless, but excess toxins over a long period of time can damage the liver e.g. alcohol)
- stores vitamins (A, D, K and B$_{12}$) and minerals (Fe and K)
- deaminates excess amino acids (converted to urea and excreted through the kidneys)

Hepatic vein

Unwanted amino acids

UREA

STORED
Glycogen
Minerals
Vitamins

Hepatic artery

Liver

Ileum

Hepatic portal vein
(rich in absorbed food)

- produces blood proteins (e.g. fibrinogen, needed for clotting)
- controls the amount of glucose in the blood (if blood sugar levels fall, insulin is produced and glycogen is converted to glucose)
- converts excess carbohydrates to fat for storage in the cells beneath the skin.

locus	(Plural = loci.) The position of a gene on a chromosome.

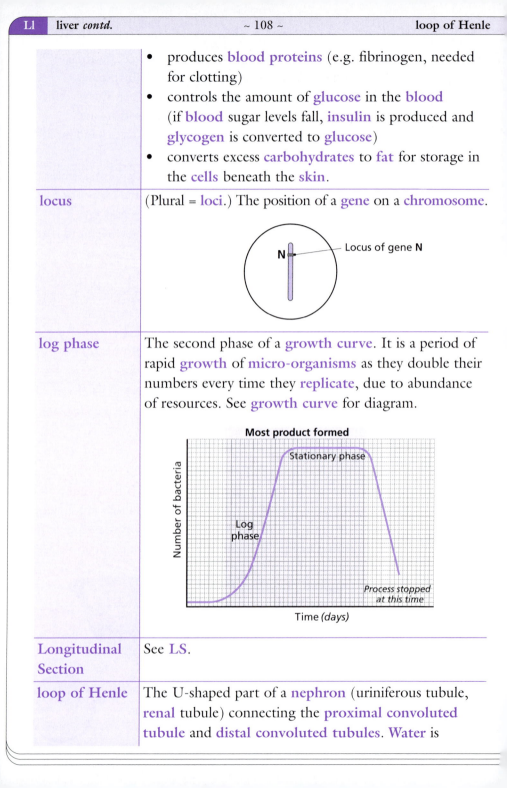

log phase	The second phase of a growth curve. It is a period of rapid growth of micro-organisms as they double their numbers every time they replicate, due to abundance of resources. See growth curve for diagram.

Most product formed

Number of bacteria

Stationary phase

Log phase

Process stopped at this time

Time *(days)*

Longitudinal Section	See LS.
loop of Henle	The U-shaped part of a nephron (uriniferous tubule, renal tubule) connecting the proximal convoluted tubule and distal convoluted tubules. Water is

reabsorbed here during the formation of urine.
See nephron for diagram.

LS	**Longitudinal Section**. View of a section of an organism cut along the long axis e.g. as in earthworm cut from head to anus.
lung	One of a pair of spongy organs that functions during breathing for gaseous exchange. See respiratory system for diagram.
lymph	Excess extracellular fluid returns to the blood system through the capillaries or drains into lymph vessels (lymphatics – thin walled with valves and lymph nodes). Fluid now known as lymph, contains lymphocytes (white blood cells).

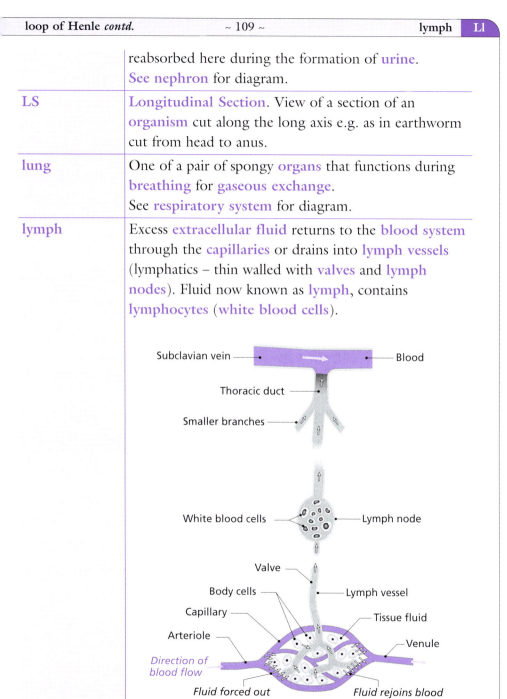

lymph node	Produces **lymphocytes** and **antibodies** that destroy invading **pathogens**. See **lymph** for diagram.
lymphatic system	A collection of fluid filled **vessels** and **organs** which: • return **tissue** fluid (**extracellular fluid**) to the bloodstream • filter **lymph** • produce leucocytes and **antibodies**.
lymphatic vessel	Tubules carrying **lymph** which drain into the thoracic duct and right lymphatic duct, which empty into the left and right sub-clavian **veins** respectively. See **lymph** for diagram.
lymphocytes	These are **white blood cells** that are formed in the **bone marrow** and bring about the **immune response**. They include **T-cells/lymphocytes** and **B-cells**.
lysis	Disintegration of bacterial and other **cells** by rupturing of the cell wall or membrane and allowing the **cell** contents to escape.
lysogenic cells	**Cells** into which the **DNA** of a **virus** has been incorporated, with no apparent disadvantageous side-effects.
lysozyme	A **protein**-digesting **enzyme** (proteolytic) that breaks down **cell membranes**.

Mm

macro-molecules	These are large molecules made up of many smaller organic molecules e.g. carbohydrates, lipids, proteins, and nucleic acids.
macro-nutrient(s)	Nutrient(s) needed by plants in fairly large amounts for successful development e.g. nitrogen, phosphorous, potassium, calcium, magnesium, sulphur and iron. Compare micronutrients.
male gamete nuclei	Two nuclei found in pollen tube as it approaches the micropyle of the ovule. One will fertilise the egg and the other will join with the two polar nuclei to form the triploid (3n) endosperm nucleus. See double fertilisation for diagram.
male reproductive system	The organs involved in gamete formation and sexual intercourse in the male.

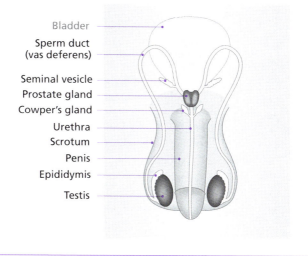

Bladder
Sperm duct (vas deferens)
Seminal vesicle
Prostate gland
Cowper's gland
Urethra
Scrotum
Penis
Epididymis
Testis

malignant	Usually refers to a tumour that is cancerous or has a tendency to spread to a different part of the body, or one that recurs after it has been removed.

malpighian layer	(Of skin.) Innermost/lowest layer of epidermis. Cells here are actively dividing (mitosis) and contain melanin (pigment responsible for tanned/dark skin). See skin for diagram.
maltase	Enzyme produced by the small intestine that converts maltose to glucose during the process of digestion.
maltose	A disaccharide sugar, $C_{12}H_{22}O_{11}$, formed from the hydrolysis of starch.
mammal(s)	Animals with hair and mammary glands that produce milk for feeding their young (see breastfeeding). Are also warm-blooded, have a four-chambered heart and have lungs.
marrow	Found in long bones, produces blood cells and contains fat-storage tissue. See spongy bone. See bone for diagram.
mechano-receptors	Sensory receptors that respond to physical changes. See receptor.
medulla	Central part of kidney (inside the cortex) containing the pyramids.
medulla oblongata	Part of the brain directly above the spinal cord responsible for involuntary muscle activities e.g. breathing, heartbeat, saliva production and swallowing. See the brain for diagram.
megaspore mother cell	The cell that undergoes meiosis to produce the megaspore. Also called the embryo sac mother cell. Compare microspore mother cell. See ovule for diagram.
megaspores	These are four haploid cells produced by meiosis in the ovule of a flower. Three of these cells will degenerate and the remaining cell nucleus divides three times by mitosis to produce eight haploid nuclei. These arrange themselves as follows: three at the top, three at the bottom and two polar nuclei in the centre. The one in the middle on the bottom is the egg cell (n).

meiosis	A **cell division** of **gamete** or **spore** producing **cells**. • Can take place in **diploid cells** only. • Produces four **cells** that may all be different genetically. • Daughter **cells** have half the number of **chromosomes** as the parent **cell**. • **Meiosis** helps to maintain the **chromosome** number of the parent by producing **haploid gametes** in **sexual reproduction**. • It also introduces **variations** into a **species** by mixing the genetic material that passes on to the next generation. 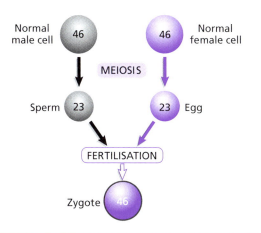
membrane	Thin layer of **tissue** that covers, connects or lines various **organs** or structures.
memory T-cells	**Lymphocytes** that are long-lived in the **circulatory system**. Produced when **T-lymphocytes** are stimulated. They memorise the **immunity** and provide a rapid response for successive exposures to an **antigen**.
Mendel's 1st Law	**Law of Segregation**. States that **characteristics** (**traits**) are controlled by pairs of **genes** (e.g. Tt) which separate (segregate) at **gamete** formation. Each **gamete** carries only one **gene** for the **trait**. See the **law of segregation** for diagram.

Mendel's 2nd Law	**Law of Independent Assortment**. States that during **gamete** formation each member of a pair of **genes** may combine randomly with either of another pair. See the **law of independent assortment** for diagram.
meninges	**Membranes** surrounding the **brain** and **spinal cord**. See the **brain** for diagram.
menopause	Period in a woman's life usually between the ages of 40 and 50 during which **menstruation** stops.
menstrual cycle	A series of changes undergone by the **uterus** in preparation for receiving a fertilised **egg**. Takes about 28 days, controlled by **hormones**. See **menstruation**.
menstruation	The discharge of menstrual fluid consisting of **blood**, lining of womb (**endometrium**) and unfertilised **egg**, which occurs monthly from **puberty** to **menopause**.
meristem	Tip of **shoots** and **roots** of plants. Area of **active cell division** (**mitosis**) which produces 'simple' **cells** which later undergo elongation and **differentiation** to give rise to the various plant **tissues** e.g. **xylem**, **phloem**, etc.

mesentery	Two layers of peritoneum enclosing neurons, blood vessels and lymph vessels which attach and support parts of the intestines to each other and to the back wall of the abdomen.
mesoderm	The middle layer of the three primary germ layers of a triploblastic organism, from which muscles, skeleton, kidneys, etc. develop. See germ layer for diagram.
messenger RNA	See mRNA
metabolic reaction	See metabolism.
metabolism	All chemical processes in living cells whereby nutritive material is either built up into protoplasm (constructive metabolism, anabolism, assimilation) or protoplasm is broken down into simpler substances (destructive metabolism, catabolism, digestion). Enzymes catalyse metabolic reactions.
metacarpals	The bones that make up the palm of the hand. See the arm for diagram.
metaphase	Stage during mitosis and meiosis during which the chromosomes (= one pair of chromatids) become attached to the spindle fibres by their centromeres and become aligned in an equatorial plane perpendicular to the spindle axis.

Two fibres attached to each chromosome

Pole

Pole

Chromosomes lined up

metatarsals	The **bones** that make up the foot between the ankles and toes. See the **leg** for diagram.
microbe	A **micro-organism** especially a **disease** causing **bacterium**.
micronutrient	See **trace element**. Compare **macronutrient**.
micro-organism	**Organism**, usually **bacterium** or **virus**, not visible to the naked **eye**.
micro-propagation	The growing of a new plant from a small piece of **stem**, **leaf** or **root tissue** in a container of **sterile nutrient** medium which contains **hormones** and **growth** substances.

Mature carrot plant

A single cell or small number of cells from carrot

Young plant

1

Nutrient (culture) medium

2

5

3

4

Callus

Callus

Different culture medium

micropyle	A small opening found between the **integuments** of an **ovule** through which a **pollen** tube can gain access to the **embryo sac** to deliver its **nuclei** and effect **fertilisation**. See **ovule** for diagram.

microscope   Apparatus used to produce enlarged images of thin/small objects.

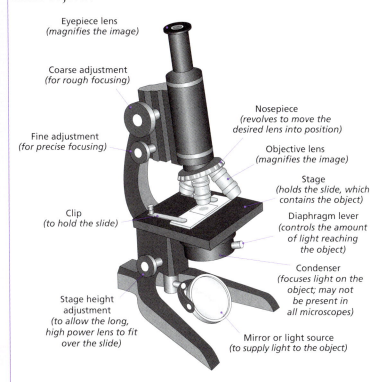

Eyepiece lens
(magnifies the image)

Coarse adjustment
(for rough focusing)

Nosepiece
(revolves to move the desired lens into position)

Fine adjustment
(for precise focusing)

Objective lens
(magnifies the image)

Stage
(holds the slide, which contains the object)

Clip
(to hold the slide)

Diaphragm lever
(controls the amount of light reaching the object)

Condenser
(focuses light on the object; may not be present in all microscopes)

Stage height adjustment
(to allow the long, high power lens to fit over the slide)

Mirror or light source
(to supply light to the object)

microspore mother cell   Cells in the pollen sacs from which microspores are produced. In flowering plants, the microspore is the pollen grain, and contains three nuclei. Compare megaspore mother cell. See pollen mother cell.

microspores   Pollen grains. Each pollen mother cell, in the pollen sacs of the anther, divides by meiosis. These are the microspores which later become the pollen grains, enclosed by a thick wall. Nucleus divides by mitosis to produce a tube nucleus and a generative nucleus. After pollination the generative nucleus divides again by mitosis to produce two male gametes – both haploid (n). See pollen mother cell for diagrams.

microvilli	Minute finger-like projections on cells e.g. on the cells of the proximal convoluted tubule of the nephron.
middle lamella	Thin layer or plate-like structure separating two plant cells. The cell plate laid down during telophase of mitosis forms the middle lamella.
migration	Move from one habitat to another – may be seasonal.
milk	Opaque white fluid secreted by the mammary glands of female mammals and used for nourishment of their young. See breastfeeding.
milk tooth	Temporary tooth or one of the first set of teeth in mammals.
mineral (element)	Inorganic compound needed in small quantities for the correct functioning of the body, or an organism. Some form part of biomolecules. Those that do not include sodium, chlorine, potassium and calcium – found as dissolved salts. Iron, copper and zinc are also important. Minerals are used by organisms in three ways: (i) to form rigid body framework: calcium in bones and plant cell walls (ii) to form body tissues e.g. iron used to make haemoglobin for blood (animal): magnesium used to make chlorophyll for photosynthetic cells (plant) (iii) to function in cellular and body fluids.

mitochondrion	(Plural = **mitochondria**.) **Organelle** found in all **cells**. Has a double **membrane** with folds called **cristae**. Matrix medium in between. Surface of **cristae** and matrix contain **enzymes** involved in **respiration** (**Kreb's cycle** and oxidative **phosphorylation**). The number of **mitochondria** in a **cell** depend on the **cell** activity e.g. high numbers in **cells** of **muscle**, **nerves**, **liver** and in **meristems**.

Outer membrane
Inner membrane

mitosis	Usual method of **cell division** by **cells** not involved in the formation of **gametes** (**somatic cells**). Occurs in a series of stages: **prophase**, **metaphase**, **anaphase**, **telophase** and a resting or preparatory stage = **interphase**. Results in the production of two daughter **cells** genetically identical to the original in every way except size. Takes place in **haploid** and **diploid cells**. Daughter **cells** have same number of **chromosomes** as parent **cell**. **Mitosis** is a method of **reproduction** in **unicellular organisms** but is mainly for **growth** and repair in **multicellular organisms**.
mitral valve	See **bicuspid** valve.
molecule	A group of atoms; the smallest part of a **compound** that still has all the chemical properties of that **compound**.

monera	One of the five kingdoms of classification: animal kingdom, plant kingdom, fungus kingdom, protist kingdom (see protista) and monera (prokaryote) kingdom. The monera kingdom includes all the bacteria.
monocot	Abbreviation for monocotyledon. Refers to plants that are herbaceous, have embryo/seed with one cotyledon (rarely stores food, absorbs food from endosperm and passes it to the embryo), flower parts in units of three, leaves with parallel veins and a stem with scattered vascular bundles (no cambium) and fibrous roots, e.g. grasses. See dicotyledon.
monocytes	Largest of the white blood cells. Formed in the bone marrow. Amoeboid. Engulf bacteria and dead cells.
monohybrid cross	A genetic cross which examines the transmission of one trait, e.g. height, eye colour, etc. Compare dihybrid cross.
mono-saccharide(s)	See carbohydrate.
morula	A solid mass of cells resulting from a series of mitotic divisions of a fertilised egg, before the blastocyst stage.
motor neuron	Efferent neuron. Carries messages from the central nervous system (CNS) to an effector. Cell body located at end of axon, inside CNS. See neuron for diagram.
mRNA	Messenger RNA. A type of ribonucleic acid found in the nucleus (10%) and cytoplasm (90%). Called mRNA because it carries the instructions for protein manufacture from the DNA in the nucleus to the ribosomes in the cytoplasm, where protein is made.

mucus	Slimy substance produced by mucous membranes, consisting mainly of glycoprotein. Mucus lubricates, moistens and protects tissues, and may contain enzymes.
multicellular	Consisting of more than one cell or many cells.
muscle	A tissue that can only contract and relax (cannot expand or elongate). To contract it needs energy – ATP from respiration of glucose or glycogen with oxygen. Consists of long muscle cells lying side by side in bundles, surrounded by sheath of tissue that extends at ends to form tendons for attachment to bones. See synovial joint for diagram.
mutagen	A substance or agent capable of causing or bringing about a mutation.
mutation	A spontaneous change in the sequence of nitrogenous bases in a gene or chromosome. See chromosome mutation and gene mutation.
mutualism	Describes two organisms of different species, both of which benefit from a close relationship e.g. a lichen is composed of an alga and a fungus intertwined. The alga obtains support and a mineral supply from the fungus; the fungus obtains food from the alga. See symbiosis.
mycelium	(Plural = mycelia.) Collective term for all the hyphal filaments that make up the vegetative part of a fungus. See *Rhizopus* for diagram.
mycorrhiza	Symbiotic relationship between a fungus and plant roots. The mycelium forms a web over the surface of the roots (exotrophic mycorrhiza) and their hyphae enter the roots (endotrophic mycorrhiza).
myelin sheath	Lipid sheath formed by the Schwann cells of neuron. Insulation on dendrites or axons of neurons. Speeds up impulse transmission. See neuron for diagram.

myopia	Short sightedness. Cannot see objects far away from the eye. Eyeball too long. Corrected using concave lenses. Compare hyperopia.
myosin	A protein found in striated muscle.
myxoedema	Results from a deficiency of thyroxine in an adult. Symptoms include: decreased rate of metabolism, slurred speech, loss of memory, dry skin, and coarse hair.

Nn

n	See haploid.
NAD	Nicotinamide adenine dinucleotide. An oxidising and reducing co-enzyme (a non-protein substance attached to some enzymes to allow them function properly) used in respiration. It traps and transfers electrons for cell activities.
NADP	Nicotinamide adenine dinucleotide phosphate. An oxidising and reducing co-enzyme (a non-protein substance attached to some enzymes to allow them function properly) necessary in photosynthesis. It traps and transfers hydrogen ions in cell activities.
natural immunity	The ability of the body to resist infection but does not involve the production of antibodies.
natural selection	The way nature selects organisms with advantageous genes which allow them to adapt to the environment, and pass them on to successive generations.
nectary	Gland found on any flower part which secretes nectar (a sweet liquid used by bees to make honey).
negative feedback	A process which stops the synthesis of a hormone because the products of the reaction (another hormone) inhibit production of the first hormone that controlled its formation.

Pituitary gland — T.S.H.

Negative inhibition

Iodine (from diet)

Thyroid

Normal level of thyroxine

Thyroxine

Carried around the body to control metabolism

nephron	Tubule in the kidney. Is the functional unit of the kidney. 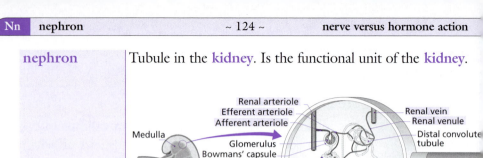
nerve cord	Solid strand of nerve fibres extending backwards from the brain e.g. spinal cord. See the brain for diagram.
nerve impulse	Chemical change transmitted along a neuron that brings about a response in an effector.
nerve versus hormone action	1. Similarities: (i) Both provide a means of communication within the body. (ii) Both involve the transmission of a message, which is triggered by a stimulus and produces a response. (iii) The target organs of a hormone are equivalent to a nerve's effector. (iv) Both involve chemical transmission. The transmission of the message across the synapse is achieved by acetylcholine (chemical messenger), which is equivalent to a hormone in the endocrine system. *contd. . . .*

2. Differences:

	ENDOCRINE	NERVOUS
1	Chemical substance sent through bloodstream	Action potential sent along **neuron**
2	Responses usually slow	Response usually rapid
3	**Hormones** carried to all parts of the body	**Impulse** transmitted to specific destinations
4	Responses often widespread – involving many **target organs**	Responses may be localised – involving only one **muscle**
5	Responses may continue over a long time	Responses usually short-lived

nervous system   A collection of nerve cells in an animal used to detect changes (stimuli) in both the internal and external environments and to coordinate a rapid response to the stimuli.

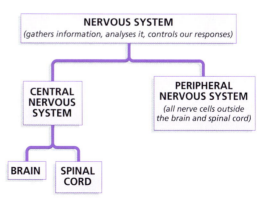

neural tube   Tube formed in embryo of vertebrates from which the brain and spinal cord develop.

neuron	Nerve cell. All neurons are basically similar in structure. Each consists of one axon, one or more dendrons, neurotransmitter vesicles (pouches) and a nucleus contained in the cell body. A myelin sheath and Schwann cells may also be present.

There are three types of neuron: sensory neurons, motor neurons and interneurons.

Sensory (afferent) neuron   Motor (efferent) neuron

neuro-transmitter	Chemical substances that carry nerve impulses across synapses. See acetylcholine.
niche	A position or place occupied.

Or

The functional role of an organism in an ecosystem, e.g. how it feeds, what it feeds on, what feeds on it, etc. |

night blindness	Where the ability to see in dim light is lessened, usually due to a deficiency of vitamin A.
nitrification	The conversion of ammonia to nitrites and nitrates. See nitrogen cycle for diagram.
nitrifying bacteria	Bacteria found in the soil. They convert ammonia (NH_3) to nitrite ($-NO_2$) then to nitrate ($-NO_3$).
nitrogen cycle	The cyclical path that nitrogen atoms pass along: from the air to compounds in the soil which are taken up by plants, then eaten by animals, and eventually return again to the air.

nitrogen fixation	The conversion of atmospheric nitrogen into a form usable by plants i.e. nitrates.
nitrogen-fixing bacteria	Bacteria that are capable of converting gaseous nitrogen from the air into nitrates ($-NO_3$). These bacteria can be (a) free-living in soil, or (b) nodules of legumes.

nitrogenous base	Nitrogen containing compound that forms nucleotides, which are the building blocks of DNA and RNA. Five types: adenine, cytosine, guanine, thymine and uracil.
nocturnal	Refers to an organism that is active at night. Compare diurnal.
node	Point on a stem where a leaf or leaves are attached. See leaf for diagram. Or Rounded structure e.g. lymph node.
node of Ranvier	Of neuron: constriction of myelin sheath on the axon to separate Schwann cells on some nerve cells. Speeds up impulse transmission. See neuron for diagram.
nodule(s)	A small rounded lump e.g. on roots of legumes.
non-cyclic photophos-phorylation	Occurs during the light stage of photosynthesis. Some electrons emitted by chlorophyll are passed onto NADP and become NADP⁻. These electrons are not returned to the chlorophyll molecule. Light causes water to split (dissociate) into hydrogen ions (H⁺) and hydroxyl ions (OH⁻). The OH⁻ being negatively charged has an extra electron and this electron is passed back to the chlorophyll molecule (hence 'non-cyclic') via an electron carrier system, which results in the formation of more ATP.

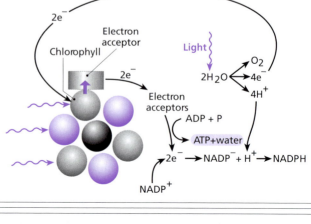

non-endospermic seed	Main food store is in cotyledons e.g. broad bean. See cotyledon for diagram.
non-nuclear inheritance	DNA is found in other organelles of the cell apart from the nucleus e.g. mitochondrial DNA and chloroplast DNA. These organelles play no part in sexual reproduction but are present in the female gametes (eggs). At fertilisation it is only the nucleus of the male gamete that fuses with the egg. All the mitochondrial DNA you possess you got from your mother, and she got hers from her mother, and so on. As a result there are very few different types of mitochondrial DNA.
notochord	A column of cells that runs along the length of the embryo, around which the vertebral column develops.
nucellus	Nutritive tissue found in the ovule and nourishes the embryo sac during its development. See ovule for diagram.
nuclear membrane	Membrane surrounding the nucleus of a cell. See cell for diagram.
nucleic acid	Two types, RNA and DNA. Long chains of nucleotides found in cytoplasm and nucleus of all cells, plant and animal.
nucleolus	Spherical body found in nucleus of non-dividing cells. Functions in protein synthesis. See nucleus for diagram.
nucleotide(s)	Made up of a phosphate group, a 5-carbon sugar deoxyribose ($C_5H_{10}O_4$) or ribose ($C_5H_{10}O_5$) and one nitrogen base: adenine (A), cytosine (C), guanine (G), thymine (T), uracil (U).

| nucleus | (Plural = nuclei.) Controls activity of cells. Contains genetic material (chromosomes) that is passed on to future generations. May contain nucleoli, which function in protein synthesis. |

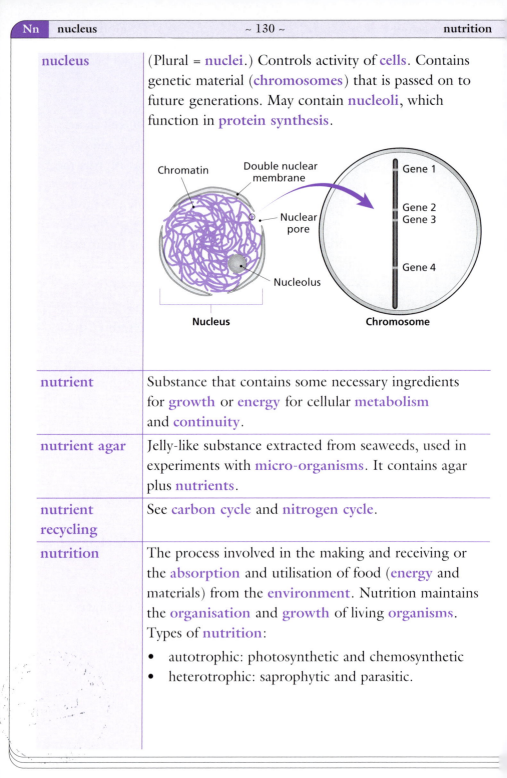

nutrient	Substance that contains some necessary ingredients for growth or energy for cellular metabolism and continuity.
nutrient agar	Jelly-like substance extracted from seaweeds, used in experiments with micro-organisms. It contains agar plus nutrients.
nutrient recycling	See carbon cycle and nitrogen cycle.
nutrition	The process involved in the making and receiving or the absorption and utilisation of food (energy and materials) from the environment. Nutrition maintains the organisation and growth of living organisms. Types of nutrition: • autotrophic: photosynthetic and chemosynthetic • heterotrophic: saprophytic and parasitic.

Oo

obligate aerobe	Organisms that only survive in the presence of oxygen.
obligate anaerobes	Organisms that only survive in absence of oxygen e.g. some bacteria.
obligate parasite	An organism (parasite) that is bound or compelled to live in another living cell and can only replicate within the cell.
oesophagus	Tube that goes from the mouth to the stomach. See the digestive system for diagram.
oestrogen	A female sex hormone produced by the ovaries. • Stimulates proliferation of uterine wall (endometrium). • Inhibits FSH production. • Stimulates LH production. • Maintenance of the female secondary sexual characteristics.
offspring	Young born to animals; children, descendants.
oil	Has the same basic structure as a fat, but contains different fatty acids. Oils are a type of lipid that is liquid at room temperature. See fat.
omnivore	An omnivorous animal i.e. an animal that feeds on plants and animal flesh. Compare herbivore and carnivore.
oncogene	A gene that causes cancer.

open circulation	Refers to a transport system that does not confine the blood to a collection of tubules e.g. insects. The blood leaves the tubules and flows among the body cells.

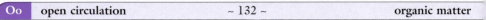

Opening in heart

Tubular heart

Blood flows around body cells

Body cells

Collecting vessels

Distributing vessels

Muscular heart

optic nerve	Nerve that transmits messages from eye to brain. See the eye for diagram.
optimum conditions	Best or most favourable conditions for growth, reproduction, enzyme activity, etc.
organ	A structure containing a group of tissues with common function(s) e.g. plant organs: leaf, root, stem, flower; animal organs: stomach, ear, testis, ovary, etc.
organ systems	A group of organs working together to carry out a function. See system.
organelle	A structure in a cell specialised to perform a particular function e.g. mitochondrion, chloroplast, etc. A membrane surrounds most organelles.
organic matter	Dead or decaying plant and animal remains. Material containing the element carbon.

organic molecule	A molecule based on carbon-hydrogen bonds, from simple compounds such as CH_4 to highly complex molecules such as proteins. Compare inorganic molecule.
organisation	Organisms are composed of cells. These cells function together to form tissues, organs, organ systems, individuals (organisms) and populations.
organism	Living thing or group of systems functioning together for living e.g. human, plant, mouse, etc.
osmoregulation	Controlling the osmotic pressure within an organism by regulating the amounts of salt and water present. It's a function of the kidneys in humans and the contractile vacuole in *Amoeba*.
osmosis	A special case of diffusion. It's the movement of solvent (always water) from a region of high solvent concentration to a region of lower solvent concentration through a semi-permeable membrane. No energy used by the cell for osmosis to take place i.e. it is a passive process. Examples of osmosis: water entering root hair cell; water moving from cell to cell in transpiration. See diffusion.

Semi-permeable membrane

Pure water		Salt water
High water concentration		Low water concentration
Low solute concentration		High solute concentration
○ = water		◯ = salt

Direction of water movement
*due to **osmosis***

ossicles	Small bones in middle ear (malleus = hammer, incus = anvil, stapes = stirrup). These amplify vibrations of eardrum. See the ear for diagram.

ossification	The process of replacing cartilage with bone in the skeleton.
osteoarthritis	A degenerative disease of joints. Affected joints have a rough surface with loss of hyaline cartilage. The structural elements of the joint begin to proliferate and the joint becomes remodelled. Surrounding tissues also become enlarged. Affected people experience gradual onset of pain worsened by exercise, and morning stiffness. As the disease progresses, joint mobility is reduced. When physical activity is severely restricted, the joint can be replaced.
osteoblast	Bone-forming cell found lining the outer surface of all bones. Also present inside most bone cavities. These cells secrete a very strong bone matrix made mainly of collagen fibres, which provide bone with its strength. They replace cartilage with bone when the body is growing.
osteoclast	A large cell, having more than one nucleus, that can break down and absorb calcified bone i.e. remove worn cells and form the central cavity in a long bone and deposit calcium into the blood.
osteocyte	A bone cell found embedded within the bone matrix. An osteoblast that has ceased its function of forming bone.
osteoporosis	A reduction in the density of bones as a result of the ageing process or from enforced inactivity. It is caused by excess reabsorption of bone and leads to an increased risk of fractures. Osteoporosis frequently follows the menopause or may be induced by long-term treatment with steroids, and may occur in males as well as females. Diagnosis is normally made using a DEXA scan (dual-energy X-ray absorptiometry).

oval window   *Fenestra ovalis.* A membrane covered opening leading into the inner ear. It passes vibrations into the cochlea.

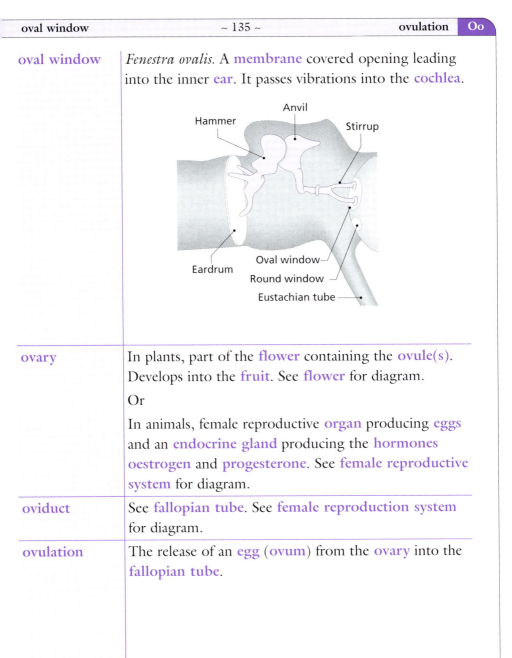

ovary   In plants, part of the flower containing the ovule(s). Develops into the fruit. See flower for diagram.

Or

In animals, female reproductive organ producing eggs and an endocrine gland producing the hormones oestrogen and progesterone. See female reproductive system for diagram.

oviduct   See fallopian tube. See female reproduction system for diagram.

ovulation   The release of an egg (ovum) from the ovary into the fallopian tube.

ovule

In ovary of flower, consists of nucellus, embryo sac and integuments. When fertilised develops into the seed.

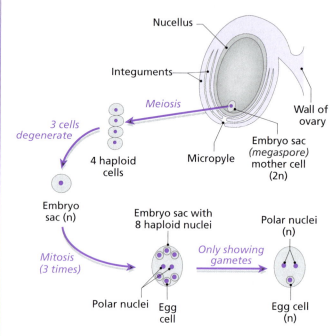

ovum

See egg.

oxidation

A chemical reaction in which oxygen combines with another substance or

a reaction which involves the removal of hydrogen from a compound or

a reaction that involves the loss of electrons.

oxygen

Symbol = O. Formula = O_2. Needed by all cells to release energy from food in respiration.

$$C_6H_{12}O_6 + 6O_2 \rightarrow energy + 6CO_2 + 6H_2O$$

It is a waste product of photosynthesis. It is taken into the body through the lungs and transported around the body in the form of oxyhaemoglobin. Oxygen is also necessary for (a) germination of seed, and (b) in the soil allows the root hair cells to respire.

oxygen debt	The amount of oxygen in a cell is kept fairly constant. During physical exercise the demand for oxygen by the muscles exceeds the supply: this results in oxygen debt. When exercise is complete the consumption of oxygen remains above normal until the debt has been repaid. This is why one continues to pant after exercise.
oxyhaemo-globin	Hb_4O_8. Oxygenated haemoglobin i.e. haemoglobin with oxygen attached. Formed in the red blood corpuscles (no nucleus) as blood passes through the lungs.
oxytocin	A hormone produced by the posterior lobe of the pituitary gland in response to the sudden fall in the levels of oestrogen and progesterone, oxytocin stimulates contractions of the uterus during labour.
ozone	A triatomic molecule of oxygen i.e. O_3. Formed when sunlight strikes oxygen atoms. Is found as a layer of the atmosphere about 30 km above the earth and prevents most of the ultraviolet radiation from the sun reaching the earth.

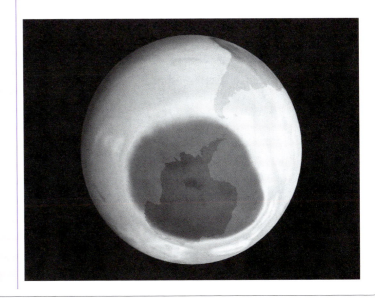

Pp

pacemaker	(S-A node) A tissue in the heart that controls the heartbeat. Also called the sinoatrial node, located in the right atrium. Sends out wave of impulses to muscles of both atria causing atria to contract. The impulses are picked up by the atrioventricular node (A-V node) and transmitted to the ventricles via the Bundle of His, causing the ventricles to contract.
pancreas	Large exocrine gland under the stomach that secretes enzymes (trypsinogen, amylase and lipase), which enter the duodenum via the pancreatic duct. It is also an endocrine gland that secretes insulin. See digestive system for diagram.
pancreatic juice	Liquid produced by the pancreas composed of enzymes (trypsin, amylase and lipase) which have a role to play in digestion.
parasite	Organism that lives in or on another organism (the host) from which it derives its nourishment and causes it harm. See endoparasite and ectoparasite.
parasitic bacteria	Bacteria that cause disease. They live in or on living organic matter.
parasitism	Association involving a parasite.
parathyroid	Endocrine glands found behind the thyroid gland in the neck. These glands secrete a hormone (parathromone) which regulates calcium reabsorption from the bones to the blood. See endocrine gland for diagram.
Parkinson's disease	Progressively degenerative disease of the nervous system due to the lack of the neurotransmitter dopamine, which regulates the nerves controlling muscle activity. Results in trembling of the muscles, stiff joints and slow walk. May also effect speech and facial expressions.

parthenocarpy	Is the development of fruits without fertilisation. Results in fruits without seeds.
parturition	The process of giving birth to young, occurs in three stages: • Stage 1: lasts for 6–18 hours. Contractions begin and increase in frequency and intensity. Cervix dilates, birth canal widens, amnion ruptures = breaking of the waters. • Stage 2: lasts for 20–60 minutes. Baby is born, mother has to 'push', baby's head appears (most difficult part now over) then shoulders, one at a time and finally rest of body – baby still attached to mother by umbilical cord and placenta. Baby cries with first breath and umbilical cord is cut when it stops pulsating. • Stage 3: lasts for 5–10 minutes. Delivery of placenta, membranes that surrounded the foetus and the remains of the umbilical cord (= afterbirth).

Cervix
Placenta Foetus stretches open

Foetus is born
head first

Placenta Umbilical
detaches cord Tied

The birth begins   The baby is born   The afterbirth is expelled

passive immunity	Recipient receives antiserum, containing antibodies, from an already immunised individual. This confers short-term immunity on that individual e.g. tetanus and newborn babies. Compare active immunity.
passive transport	Movement of substances across the cell membrane which does not involve the use of energy by the cell e.g. diffusion and osmosis.

pasteurisation	A method of food preservation used for milk. Temperature raised to about 72°C for 12 seconds and cooled immediately to below 10°C. Does not kill all bacteria.
pathogen	An organism that causes a disease e.g. some bacteria are capable of causing a disease. e.g. diphtheria, whooping cough and tetanus all caused by bacteria.
pectoral girdle	The bones that attach the arms to the axial skeleton i.e. the clavicle and scapula. The arms are composed of the humerus, radius, ulna, carpals, metacarpals and digits (fingers) containing phalanges. See the human skeleton for diagram.
pelvic girdle	The fused bones of the hips, attached to the sacrum surrounding a cavity, that support the legs. The legs are composed of the femur, patella, tibia, fibula, tarsals, metatarsals and digits (toes) containing phalanges.
pelvis	(Of kidney.) Central region of kidney into which urine drains from the collecting ducts. The urine is brought from here in the ureter to the bladder. See kidney for diagram.
penis	The male copulatory organ. Deposits sperm in vagina. See male reproductive system for diagram.
pentadactyl limb	Five-fingered limb found in nearly all terrestrial vertebrates. See the arm or the leg for diagram.
pepsin	An enzyme produced in the stomach when pepsinogen reacts with the hydrochloric acid. It converts protein to polypeptides during the process of digestion. Its optimum pH is acidic: pH = about 1 or 2.
peptide	Compound with two or more amino acids linked together in sequence. A very small protein. Many of these form a polypeptide chain.
peptide bond	Bond formed between peptides.

perennation	(Plants.) Surviving adverse conditions (winter) by storing food in a perennating organ (e.g. rhizome = underground stem and tap root). This food is used to produce new growth the following spring.
perennial	Plant that lives for many years and produces seed each year.
perforin	Chemical produced by killer T-cells that perforates cell membranes. Plasma or extra-cellular fluid can then enter the cells and cause them to swell and burst.
pericardium	Double layered sac enclosing the heart, fluid-filled to prevent friction when the heart beats.
period	See menstruation.
peripheral nervous system (PNS)	All nerves outside the central nervous system (CNS) i.e. all nerves except the brain and spinal cord. They carry messages to and from the CNS.
peristalsis	Method of moving substances (e.g. food) through tubes (e.g. intestines) by waves of rhythmic contractions and relaxations of muscles.
permeable	Refers to a membrane, allows molecules or ions of a certain maximum size to pass through. The cell wall is fully permeable (allows all molecules to pass through irrespective of size) and the cell membrane is semi-permeable.
perspiration	Sweating. This occurs in response to warm conditions e.g. during physical exercise. Sweat glands release sweat (95% water) onto the skin surface – this absorbs heat from the body and the water evaporates with the heat, leaving the body cooler. It is important when exercising to drink water before, during and after to try and maintain the water salt balance in the body.
pest	Any troublesome, annoying, destructive organism.

pesticide(s)	Chemicals applied to plants that kill or inhibit growth of insects or other plant pests.
petal	Part of a flower, usually coloured. See flower for diagram.
petiole(s)	The stalk or stem of a leaf, supporting the leaf blade. See the leaf for diagram.
pH	This refers to the concentration of hydrogen ions (H^+) or hydroxyl ions (OH^-). The pH scale goes from 0 to 14: $$pH < 7 \rightarrow acid; \; pH = 7 \rightarrow neutral;$$ $$pH > 7 \rightarrow base/alkaline \text{ substance.}$$
phagocytes	White blood cells that can engulf and destroy viruses and bacteria by phagocytosis. Bacteria / White blood cell (phagocyte) / Pseudopodia engulf bacteria / Bacteria are digested
phagocytitic	Of or pertaining or relating to phagocytes.
phagocytosis	Eating by cells e.g. the ingesting of food by an *Amoeba*, white blood cell ingesting bacterium.
phalanges	(Singular = phalanx.) Bones of the fingers or toes. See the human skeleton for diagram.
pharynx	That part of the digestive system that connects the mouth to the oesophagus. See the respiratory system for diagram.
phenotype	Physical appearance of an individual as a result of the interaction of the genotype with the environment.

phloem	Conducting tissue in plants for transporting e.g. food and hormones. 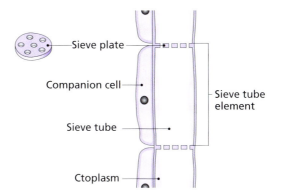 Sieve plate Companion cell Sieve tube element Sieve tube Ctoplasm
phloem sieve tube	Long cells, thin side walls, thick perforated end walls (sieve) allows passage of cytoplasm between cells. No nucleus when mature. See phloem for diagram.
phospholipid	A lipid with one of its fatty acids replaced with a phosphate group (see triglyceride). Needed for cell membrane formation. 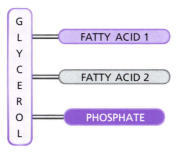
phospho-rylation	Is the addition of phosphorus (phosphate) i.e. ADP + P = ATP.
photophos-phorylation	Is phosphorylation using light i.e. the production of ATP using light during the light stage of photosynthesis.
photoreceptors	Sensory receptors that respond to the presence of light. See receptor.

photosynthesis

$$6CO_2 + 6H_2O \xrightarrow[\text{chlorophyll}]{\text{light energy}} C_6H_{12}O_6 + 6O_2$$

The process in plants that makes food using light energy. It occurs in the chloroplasts of green plants in which carbon dioxide and water in the presence of light energy and chlorophyll are converted into simple sugars and oxygen. The light energy is converted to chemical energy and is stored in the sugar molecule.

What happens to the products of photosynthesis? The simple sugars i.e. glucose are either used in respiration or converted to starch and stored in the leaf or transported in the phloem and stored elsewhere in the plant. The oxygen can be used in respiration or passed out to the atmosphere.

Chlorophyll (in chloroplasts) traps sunlight energy. This energy is used to split water (H_2O, from the soil) into protons ($2 \times H^+$), oxygen (O) and electrons ($2 \times e^-$). The electrons (e^-) are passed to chlorophyll. The protons (H^+) enter a pool of protons. Electrons from chlorophyll together with protons from the pool of protons are joined to carbon dioxide, from the atmosphere, (i.e. CO_2 is reduced) to form a carbohydrate $C_x(H_2O)_y$.

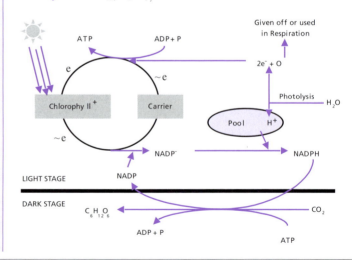

photosynthetic bacteria	Autotrophic green sulphur bacteria, get energy from sun.
phototropism	The growth response of a plant to light, caused by the higher concentration of the plant hormone IAA (indoleacetic acid, an auxin) on the darker side of the plant shoot. This promotes cell elongation on that side and the shoot grows towards the light as a result.

Meristem makes IAA IAA diffuses down shady side Cells elongate Light Cells No elongation

phytoplankton	Plant plankton.
pinna	Ear: made of cartilage, funnels sound waves into ear. See the ear for diagram.
pituitary	Master endocrine gland of the body situated at the base of the brain, produces hormones and controls other glands. See the brain for diagram.
placebo	Dummy pill used as a control in medical tests. Does not contain any 'medicine'. See also double blind testing.
placenta	A structure in the carpel of a flower that attaches the ovule to the wall of the ovary. Or Structure attached to the inner surface of the womb (uterus) of pregnant mammals, which helps to nourish the foetus, and discharges its waste. Its functions include: 1. Exchange of materials: diffusion of food, oxygen, minerals, antibodies and waste products takes place in the placenta from one bloodstream to the other.

contd. ...

Drugs, viruses and chemicals in cigarette smoke can also pass into the foetus.

2. Acts as a barrier: against blood pressure of mother, may be too great for foetus. Protects foetus from mother's immune system which may reject the foetus as foreign.

3. Endocrine gland: secretes progesterone and oestrogen. These maintain endometrium and prepare breasts for lactation. See parturition for diagram.

placenta formation	When the blastocyst reaches the uterus it sinks into the endometrium – this is implantation. The trophoblastic layer forms trophoblastic villi, which embed in the endometrium and eventually form the placenta. The placenta is formed partly from the tissues of the embryo and partly from the uterus wall.
plankton	Aquatic plants and animals that float suspended in the water – provide food for fish and whales.
plant growth regulators	Chemicals produced in the meristematic regions and transported through the vascular system of plants. They affect the rate of growth or development of plants when they are in very low concentrations. Some external factors also regulate the growth of plants e.g. light intensity, amount of daylight, temperature and gravity. See cytokinins.
plasma	Clear liquid portion of the blood, composed of 90% water in which red blood corpuscles, white blood cells and platelets are suspended together with a number of dissolved substances e.g. products of digestion glucose, amino acids, glycerol, fatty acids, minerals, vitamins, waste products (e.g. carbon dioxide, urea), hormones (e.g. insulin), plasma proteins (e.g. albumen, fibrinogen, prothrombin), antibodies (proteins which destroy pathogens), enzymes (e.g. thrombin), salts (e.g. NaCl).

plasmid	Small circular piece of DNA found in bacteria and yeast. Often contains genes bestowing drug resistance.

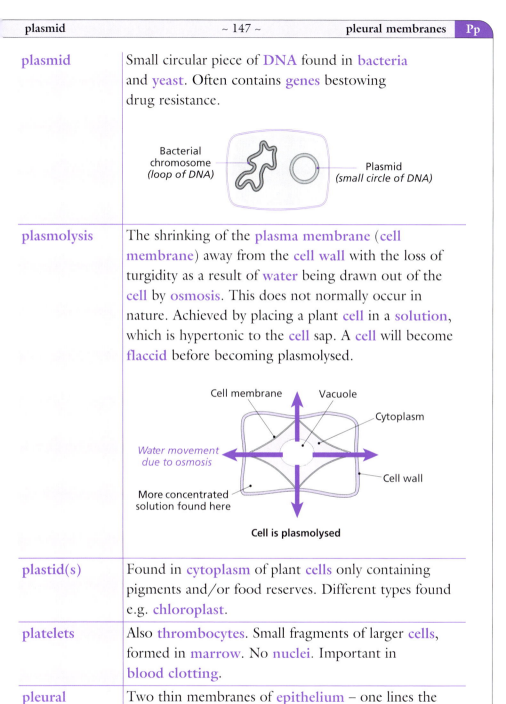

plasmolysis	The shrinking of the plasma membrane (cell membrane) away from the cell wall with the loss of turgidity as a result of water being drawn out of the cell by osmosis. This does not normally occur in nature. Achieved by placing a plant cell in a solution, which is hypertonic to the cell sap. A cell will become flaccid before becoming plasmolysed.

Cell membrane   Vacuole

Cytoplasm

Water movement due to osmosis

Cell wall

More concentrated solution found here

Cell is plasmolysed

plastid(s)	Found in cytoplasm of plant cells only containing pigments and/or food reserves. Different types found e.g. chloroplast.
platelets	Also thrombocytes. Small fragments of larger cells, formed in marrow. No nuclei. Important in blood clotting.
pleural membranes	Two thin membranes of epithelium – one lines the cavity of the thorax and the other covers the outer surface of the lungs.

plumule	Terminal bud or epicotyl of embryo in seed plants. See endosperm for diagram.
polar nucleus	One of two nuclei in the centre of the embryo sac which, when fertilised (two polar nuclei plus a male gamete nucleus from the pollen grain join together) will develop into the triploid endosperm nucleus. See embryo sac for diagram.
pollen	Collective term for the pollen grains of seed plants. See flower for diagram.
pollen grain	Microspores of seed plants produced by a pollen mother cell in the pollen sac. Contains two nuclei – tube and generative. Produces the male gametes.

Diploid cells — Meiosis — Haploid nucleus — Mitosis — Male gametes
Anther — Young pollen grain — Mature pollen grain — Exine — Intine

pollen mother cell	Cell in pollen sac from which pollen grains develop. Compare megaspore mother cell.

Filament (containing vascular bundle)
Point at which anther splits to release pollen
Anther
Filament
Pollen sac
Epidermis
Fibrous layer
Tapetum
Pollen
Pollen mother cells (diploid)
Meiosis — Tetrad of pollen — Seperates — Pollen grains

pollen sacs	Elongated or tubular containers found in the anther at the top of the stamen in which pollen grains are produced. See pollen mother cell for diagram.
pollen tube	Slender tube formed on germination of pollen grain; carries the two sperm nuclei to the opening of the embryo sac at the micropyle.
pollination	Is the transfer of pollen from the anther of the stamen of one flower to the stigma of the carpel of (a) the same flower or another flower on the same plant (self-pollination). Or (b) the transfer of pollen to another flower on a different plant of the same species (cross pollination). Wind and insects are agents of pollination.

This diagram shows a wind-pollinated flower.

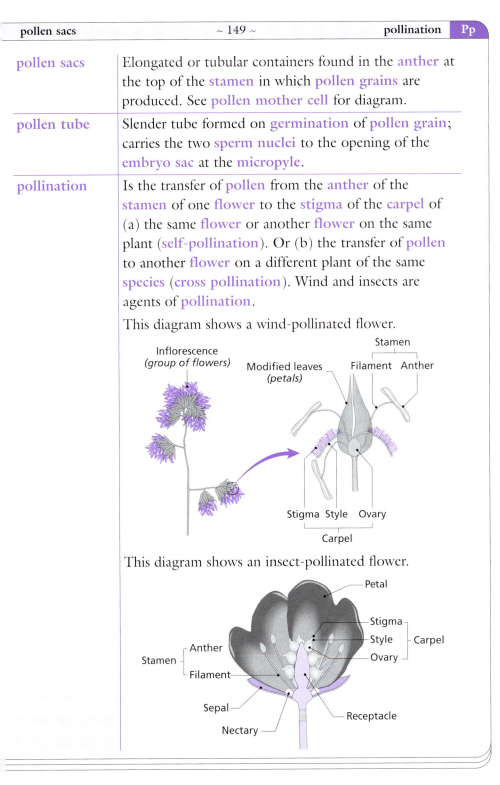

This diagram shows an insect-pollinated flower.

pollutant	Any substance, including chemicals of human origin, that contaminates a habitat or environment.
pollution	Contamination of a habitat or environment. Any human addition to the environment that leaves it less able to sustain life. It is the most harmful human impact and affects air, fresh water, sea and land.
polypeptide	Forms the basis of a protein. See peptide.
polysaccharide	Composed of many sugar units. See carbohydrate.
population	A group of organisms of the one species living in part of an ecosystem.
population curve	Graph showing the change in a population of a species over time.
population dynamics	The study of the changes in population size and the reasons or factors that bring about the changes e.g. food availability, concealment, emigration, immigration.
portal blood system	A circulatory system in which capillaries drain into a vein that opens into another capillary network i.e. it begins and ends in capillaries. e.g. hepatic portal system : capillaries (in stomach and intestines) → venules → hepatic portal vein → venules → capillaries (in the liver).

predation	The act, of some animals (predators), of capturing and killing other animals for food.
predator	Animal that hunts, captures and kills other animals (prey) for food. Have evolved adaptive techniques to survive e.g. wolf has keen hearing and eyesight, strong muscles, sharp teeth, camouflage and hunts in packs.
pregnancy	Of female animal carrying developing young in uterus. See gestation period.
preservation (of food)	Methods used to prevent food from decomposition, fermentation or deterioration e.g. salting and sugaring.
pressure filtration	Process which separates particles in a solution by molecular size, e.g. in the nephron of the kidney – water, urea, uric acid, glucose, amino acids, vitamins and mineral salts (all small molecules), filter from the glomerulus into the capsular space. Plasma proteins (albumen, fibrinogen, prothrombin – all large molecules), and blood cells will not pass through. See Bowman's capsule for diagram.
prey	An animal that is hunted and killed by another animal (predator) for food. Has evolved adaptive techniques to survive e.g. deer – keen hearing and eyesight, quick to turn and run and camouflage to evade predators.
primary consumer	A herbivore which obtains its nutrition directly from plants; usually the second member of a food chain e.g. primary producer → primary consumer → secondary consumer → tertiary consumer.
primary meristem	Tip of the main (first in terms of development) shoots and roots of plants. Area of active cell division (mitosis) which produces 'simple' cells which later undergo elongation and differentiation to give rise to the various plant tissues e.g. xylem, phloem, etc.
primary producer	First member of a food chain. See producer.

primary sexual characteristics	Distinguishing features of males and females at their birth.
producer	Any organism capable of making its own food from inorganic materials e.g. green plants. See primary producer.
progesterone	A hormone produced by the ovaries. Stimulates endometrium growth in preparation for and during pregnancy. See bacterium for diagram.
prokaryote	Cell that does not have a membrane-bound (true) nucleus and organelles e.g. bacteria. Compare eukaryote.
prolactin	A hormone produced by the anterior lobe of the pituitary, which stimulates milk production in mammary glands (breasts) and inhibits FSH production by the pituitary.
propagation	Spreading, multiplying, increasing in numbers, breeding.
prophase	The first stage of cell division (mitosis and meiosis). Chromatin threads shorten and thicken to become visible as chromosomes. These duplicate = chromatids (held together at centromere). Nuclear membrane disappears and spindle fibres form. This is followed by metaphase.

Spindle fibres form

Chromosomes become visible

Nuclear membrane breaking down

Nucleolus breaks down

proprioceptors	Sensory receptors that respond to changes in position. See receptor.

prostate gland	Involved in the manufacture of seminal fluid and provides nourishment and medium for sperm to swim. See male reproductive system for diagram.
protease	An enzyme that breaks down proteins e.g. pepsin, trypsin.
protein	Fibrous substance found in skin and hair (keratin), muscle (myosin) and various internal organs. Compound composed of the elements carbon, hydrogen, oxygen, nitrogen and sometimes sulphur. Basic unit of protein is the amino acid. There are approximately twenty different amino acids from which proteins are made. Needed by animals in their diet for growth of new cells and repair of old damaged ones.
protein synthesis	Is the function of the gene. The genes manufacture proteins (enzymes) that express themselves as visible traits. The process proceeds as follows:

- Enzymes unwind the double helix.
- A sequence of nitrogenous bases on the DNA contains the code for a particular protein. The enzyme RNA polymerase copies this sequence to form a mRNA (transcription) strand that that is complementary to the DNA strand.

Protein grows out of ribosome

Amino acids are bonded together

Amino acid separates from tRNA

tRNA

AA 6
AA 6
AA 12
AA 7

CUU

UUU

CUU GUA

AUG GAA GAA CAU AA AA

Start codon   Ribosome   mRNA   Stop codon

contd. ...

	• This **mRNA** can leave the **nucleus** and attach itself to a **ribosome**. • Each sequence of three **bases** (triplet) on the **mRNA** is a **genetic code** or **codon** that codes for a starting **codon**, a single **amino acid** or a stop **codon**. • To the **ribosome**, free-floating **tRNA** brings **amino acids** from the **cytoplasm** and lays them down in the order determined by the sequence of **bases** on the **mRNA** (**translation**). The **amino acids** bond and in this way a polypeptide chain is built up. • The **tRNAs** continue to bring **amino acids** until a stop **codon** is reached. • The **protein** then folds itself into its functional shape.
protista	(**Taxonomy.**) A large group of **organisms** of simple biological **organisation**, possess a true **nucleus** and **chromosomes** e.g. **algae**, **bacteria**, **fungi** and protozoa.
protoplasm	The entire contents of a **cell** including the **cell membrane**. **protoplasm** = **cytoplasm** + **nucleus** + **cell membrane**. See **cell** for diagram.
proximal convoluted tubule	One of two highly coiled tubules in the **nephron** of the **kidney**, this one situated nearest to **Bowman's capsule**. Reabsorption of substances needed by the body occurs here i.e. high threshold substances. See **nephron** for diagram.
pseudopodium	False foot, temporary protrusion from **cell** of *Amoeba*. See *Amoeba* for diagram.
puberty	Becoming capable of **reproduction** through natural development of the reproductive system.
pulmonary artery	**Blood vessel** that brings **blood** from the **heart** to the **lungs**. It is the only **artery** to carry deoxygenated **blood**. See the **heart** for diagram.

pulmonary circulation	The pathway of the blood from the heart to the lungs and back to the heart again. Compare systemic circulation. 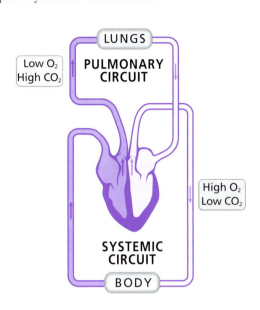
pulmonary vein	Blood vessel that brings blood to the heart from the lungs. It is the only vein to carry oxygenated blood. See the heart for diagram.
pulse	Rhythmical throbbing of arteries as a wave of pressure passes rapidly along them, resulting from the beat of the heart.
pupil	Hole in centre of the iris through which light enters the eye. See the eye for diagram.
pure-breeding	Where all the offspring from an individual (homozygous) reproducing asexually (e.g. *Amoeba*) are identical. Or Identical offspring from plants (homozygous) that self-fertilise.

purine	A group of nitrogenous bases that form part of a nucleotide. The purine bases are adenine and guanine. Compare pyrimidine.
Purkinje fibres	Special muscle fibres in the heart that conduct impulses and bring about a heartbeat.
pyloric sphincter	Ring of muscle found at the junction of the stomach and small intestine. It controls the (rate of) movement of chyme into the small intestine. See sphincter muscle. See the bile duct for diagram.
pyramid	Part of the kidney containing collecting ducts. See the kidney for diagram.
pyramid of biomass	(Ecology.) Pyramid (square base with four triangular sides sloping to apex) representing the mass of living matter or dry matter at each trophic level.
pyramid of numbers	(Ecology.) Pyramid (square base with four triangular sides sloping to apex) representing a food chain. Bottom layer is the largest and represents a very large number of primary producers; next layer smaller and represents a smaller number of primary consumers; and so on to the uppermost layer where there may be only one tertiary consumer.

It is a diagram, in block form, showing the number of individuals at each trophic level in a food chain, the size of each block indicates the number of individuals. |

Normal pyramid of numbers   Inverted pyramid of numbers

pyrimidine	A group of nitrogenous bases that form part of a nucleotide. The pyrimidine bases are cytosine and thymine in DNA and uracil in RNA, replacing thymine. Compare purine.
pyruvic acid	Pyruvate ($CH_3COCOOH$) = a 3-carbon compound formed in glycolysis. In anaerobic conditions it is converted to ethyl alcohol and carbon dioxide (in yeast) or it is converted to lactic acid in animal muscle (causes cramp). Under aerobic conditions it goes into Kreb's Cycle and is completely broken down to carbon dioxide and water – a lot of energy is released during this process.

Qq

qualitative study	A study determining the presence/absence of a substance or **organism** in a sample or **habitat**.
quantitative study	A study determining the amount of a substance or the number of an **organism** present in a sample or **habitat**.

Rr

radicle	Part of a plant embryo that develops into the root.
radius	One of the two bones in the lower arm of humans lying beside the ulna. See the arm for diagram.
reabsorption	Taking in again. Examples: • in the nephron of the kidney glucose, amino acids, etc. are taken back into the blood in the proximal convoluted tubule after being removed from it in the glomerulus • reabsorption of bone resulting in osteoporosis.
receptacle	The tips of a flower stalk (pedicel) on or around which the flower parts develop. See flower for diagram.
receptor	A structure that receives and transmits a stimulus; touch, smell, taste, pain, etc. See chemoreceptors, mechanoreceptors, proprioceptors, photoreceptors and thermoreceptors.
recessive allele	Gene which can only be expressed when both alleles are the same (homozygous condition) e.g. tt = dwarf; t is recessive.
recombinant DNA molecules	These are new combinations of DNA formed by cutting segments of DNA containing the known, wanted genes from two different sources, using restriction enzymes and then joining them together to create a new sequence of genes. Transferring genes between species is usually done using a vector e.g. a virus or plasmid.
recombinant DNA technology	This is a series of techniques in which DNA fragments containing the appropriate genes are inserted into the DNA of a host cell to create recombinant DNA molecules. The host cell then replicates these molecules to create clones of the inserted genes.

recycling	(Of nutrients in nature.) Involves the reuse of molecules as they pass from the non-living (abiotic) to the living (biotic) and back to the non-living environments continuously. Or The reuse of domestic and industrial waste.
red blood cell(s)	See erythrocyte. Membrane ——— Haemoglobin ———
reduction	A chemical reaction in which hydrogen combines with another substance. Or A reaction which involves the removal of oxygen from a compound. Or A reaction that involves the gain of electrons. Compare oxidation.
reflex action	Automatic involuntary response to an internal or external stimulus. Not under conscious control, e.g. knee jerk, ankle jerk.

reflex arc	The pathway from the point of stimulation to the responding effector. A reflex arc has five components. The stimulus is picked up by a receptor (1), that sends the message along an afferent (sensory) neuron (2), to the spinal cord (3). The reply to this stimulus is sent from the spinal cord along an efferent (motor) neuron (4), to an effector (muscle or gland) (5). 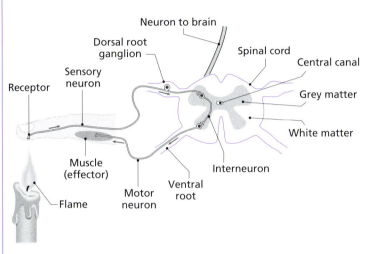
relaxin	A hormone secreted by the placenta. Relaxes the joints of the pelvic girdle and aids in the dilation of the cervix.
renal	Of or pertaining or relating to the kidney.
renal artery	Branch of the aorta bringing blood from the heart to the kidney. See the heart for diagram.
renal vein	Blood vessel bringing blood from the kidney to the inferior vena cava towards the heart. See the kidney for diagram.
replicate	Make a duplicate of, or reproduce.

replication	The process of duplication of **DNA** during **mitosis** and **meiosis**.

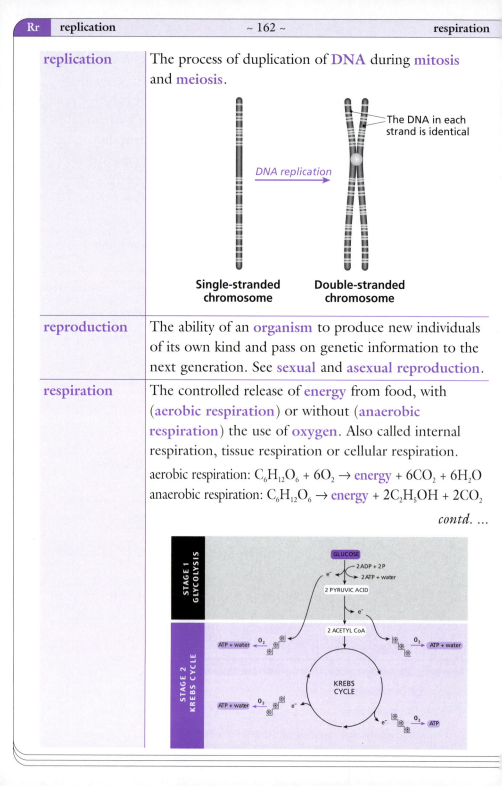

The DNA in each strand is identical

DNA replication

Single-stranded chromosome   **Double-stranded chromosome**

reproduction	The ability of an **organism** to produce new individuals of its own kind and pass on genetic information to the next generation. See **sexual** and **asexual reproduction**.
respiration	The controlled release of **energy** from food, with (**aerobic respiration**) or without (**anaerobic respiration**) the use of **oxygen**. Also called internal respiration, tissue respiration or cellular respiration.

aerobic respiration: $C_6H_{12}O_6 + 6O_2 \rightarrow$ **energy** $+ 6CO_2 + 6H_2O$

anaerobic respiration: $C_6H_{12}O_6 \rightarrow$ **energy** $+ 2C_2H_5OH + 2CO_2$

contd. ...

STAGE 1 GLYCOLYSIS

GLUCOSE

2 ADP + 2P
2 ATP + water

2 PYRUVIC ACID

2 ACETYL CoA

STAGE 2 KREBS CYCLE

ATP + water O_2

ATP + water O_2

KREBS CYCLE

ATP + water O_2

ATP O_2

Respiration may occur as a one or a two-stage process. The first stage does not require oxygen, releases only a small amount of energy and occurs in the cytosol of the cell. The second stage does require oxygen, releases a large amount of energy and occurs in the mitochondrion. Anaerobic respiration is a first stage process. Aerobic respiration is a two-stage process.

respiratory surface
Specialised region where gaseous exchange takes place, e.g. alveoli of humans, gills of fish, skin of earthworm.

respiratory system

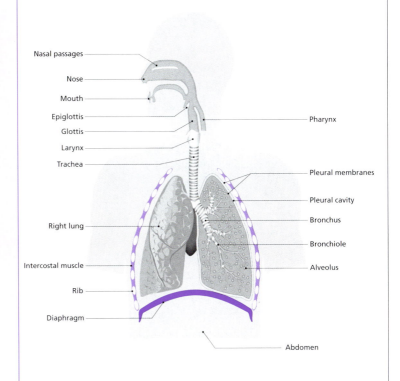

Nasal passages
Nose
Mouth
Epiglottis
Glottis
Larynx
Trachea
Pharynx
Pleural membranes
Pleural cavity
Bronchus
Right lung
Bronchiole
Intercostal muscle
Alveolus
Rib
Diaphragm
Abdomen

responsiveness
The ability of a living organism to react to changes in their internal and external environments. It is a form of defence that allows the organism to survive.

restriction enzyme	Any enzyme that recognises specific nucleotide sequences on DNA molecules and cuts the DNA strands at those sites. Some bacterial enzymes cut DNA at a specific nucleotide sequence. This is a defence mechanism used by bacteria against viral DNA and is now used as an important tool in biotechnology.
retina	Light sensitive layer at the back of the eye, contains rods (dim light) and cones (colour vision). 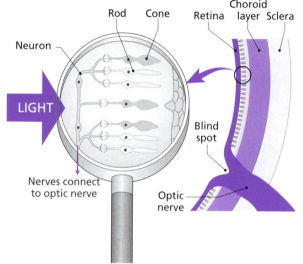
retrovirus(es)	Viruses that contain just one single strand of RNA and replicate using the enzyme reverse transcriptase to copy this RNA into a complementary single strand of DNA. The single-stranded DNA is then copied, and the resulting double-stranded DNA is inserted into a chromosome of the host cell.
reverse transcriptase	An enzyme that transcribes RNA into DNA, the reverse of transcription, used by retroviruses for replication and in biotechnology.
reverse transcription	The process of transcribing RNA into DNA (the reverse of transcription). Used by retroviruses and in biotechnology.

rhesus factor	An antigen that may be present on the surface of red blood cells. This forms the basis of the rhesus blood group system. People who have the rhesus factor are rhesus positive (Rh$^+$), while those who lack the factor are rhesus negative (Rh$^-$). The two blood types are incompatible; blood transfusions between the two types can lead to reactions.
rheumatoid arthritis	A chronic inflammation of the synovial membrane of joints. This type of arthritis is accompanied by symptoms of anorexia, fatigue and malaise. Prolonged rheumatoid arthritis may damage ligaments and lead to joint deformities. See autoimmune disease.
rhizoid	A root-like filament which attaches the mycelium of some fungi to the substrate. Grows downwards into substrate. See *Rhizopus* for diagram.
rhizome	Swollen horizontal underground stem. Perennating organ. Terminal bud produces leaves and flowers above ground. Lateral buds produce new rhizomes underground, roots (adventitious) e.g. iris, mint, scutch grass.
Rhizopus	Bread mould.

Sporangium · Spores · Columella · Apophysis · Sporangiophore · Spores blown away · Stolon · Mycelium · Hyphae · Substrate · Rhizoids

rib(s)	Curved bone(s) surrounding chest cavity, protecting heart and lungs, joined to vertebrae at proximal end. Distal end may be connected to sternum (true ribs = pairs 1 to 7) or they may be connected to other ribs (false ribs = pairs 8 to 10) or they may be free (floating ribs = pairs 11 and 12). See the human skeleton for diagram.
ribose	$C_5H_{10}O_5$. A monosaccharide found in ATP and in nucleotides of RNA. See also carbohydrate.
ribosomal RNA	See rRNA.
ribosome	Cell organelle found on endoplasmic reticulum, rich in RNA and functions in protein synthesis.
rickets	A disease due to a deficiency of vitamin D in the diet of children. Symptoms include softening of the bones of the spinal column and bow legs.
RNA	A nucleic acid found in cell nucleus (10%) and cytoplasm (90%). Involved in protein synthesis. Three types: mRNA, tRNA and rRNA.
RNA polymerase	Enzyme involved in protein synthesis – manufactures mRNA from DNA.
roan	Animal coat colour, a mixture of red and white hairs, due to incomplete dominance of genes.
rod(s)	1. (In eye.) Cells in retina of eye stimulated by low light intensity i.e. dim light – monochromatic vision (can see in black and white or shades of grey). See retina for diagram. 2. (In bacteria.) Classification grouping of bacteria called *Bacillus*. Any bacteria called *Bacillus* ~ or *B.* ~ are rod shaped; also *E. coli*, found in the human colon, is rod shaped and is responsible for the making of some B vitamins.

root	The part of a plant that grows downwards into the soil. It functions in anchorage, absorption and sometimes storage. Roots do not have leaves or buds. See root tip.
root cap	A loosely arranged mass of cells that covers and protects the root tip as it grows through the soil.

Names of parts — Zones of root

Xylem
Phloem
Root hairs — Differentiation
Epidermis
Elongation
Meristem — Cell production
Root cap — Protection

root hair	Tubular, hair-like extension of certain epidermal root cells. Found just behind apical meristem. Function is to increase surface area for absorption of water and soluble minerals. See root cap for diagram.
root pressure	The force which can push water up a stem from the root and cause it to exude from a cut stump of a plant. The movement of water into the xylem, by osmosis, causes this. This exudation is caused by root pressure.
root tip	End of root containing meristematic tissue. Area or zone of active cell division and growth. See root cap for diagram.
root tuber	Swollen fibrous roots, no eyes (buds) e.g. dahlia, lesser celandine.
roughage	Food containing a large content of indigestible material (cellulose) e.g. vegetables. See fibre.

round window	*Fenestra rotunda*. A membrane covered exit leading from the inner ear, acts as an expansion 'valve' for fluid compression in the cochlea. See oval window for diagram.
rRNA	A type of RNA found at the ribosomes, functions during protein synthesis.

Ss

saccharide	A chemical name for a sugar. See carbohydrate for the different types of saccharides.
sacrum	Region of the vertebral column consisting of five vertebrae fused together, found just above the coccyx. Hip bones are attached to the sacrum.
salinity	Refers to the amount of salt dissolved in water. The higher the salinity the more dissolved salts it contains.
saliva	Fluid produced and secreted by the salivary glands in the mouth; contains the enzyme ptyalin or salivary amylase.
salivary gland(s)	Glands that secrete saliva into the mouth. See the digestive system for diagram.
salt	A compound of a metal and a non-metal e.g. sodium chloride (NaCl), magnesium sulphate ($MgSO_4$). Formed when an acid reacts with a base. When a salt dissolves in water it breaks up into ions e.g. $NaCl \rightarrow Na^+ + Cl^-$ $MgSO_4 \rightarrow Mg^{++} + SO_4^-$
salting	A method of food preservation. Causes plasmolysis i.e. water to be removed from bacterial cells by osmosis and denatures bacterial enzymes. Kills all bacteria.
S-A node	Sinoatrial node. See pacemaker.
saprophyte	Organism that obtains its nourishment from dead or decaying organic matter e.g. bacteria and fungi.
saprophytic bacteria	Live on non-living organic matter e.g. azotobacter.
saturated fat	A triglyceride that has only single covalent bonds between the carbon atoms. Compare unsaturated fat.
scapula	Large triangular bone on the back (shoulder blade) to which the arms are joined and rotate in humans. See the human skeleton for diagram.

Schwann cell	**Cell** found around the **axon** of a **neuron** – secretes the **myelin sheath**. See **neuron** for diagram.
scientific method	A systematic approach to investigations that forms the basis for modern science: 1. Observation. 2. Forming a **hypothesis**. 3. Testing the **hypothesis** – by designing a controlled experiment. 4. Collecting and interpreting results. 5. Reaching a conclusion. 6. Reporting and publishing results. 7. Development of a theory and principle.
scion	**Shoot** of a plant specially cut for grafting onto **stock**.
sclera	The white outer coat/layer of the **eye**. Maintains eyeball shape, strong, opaque. Provides **muscle** attachment. See the **eye** for diagram.
scramble competition	Is where each **organism** tries to acquire as much of the resource as possible e.g. an ivy plant and a hawthorn tree may compete for light. The ivy uses **adventitious roots** to grip the hawthorn and climb higher.
scrotum	External pouch in most male **mammals** that contains (holds) the **testes** outside the body at a temperature slightly lower (35°C) than body temperature (37°C in humans) for the efficient production of **sperm**. See **male reproductive system** for diagram.
sebaceous gland	Produces an oily substance called **sebum**. Keeps hair supple. See the **skin** for diagram.
sebum	Oily substance produced by **sebaceous glands**. Keeps hair supple.
secondary consumer	A **carnivore** which obtains its **nutrition** from animals, usually the third member of a **food chain** e.g. **primary producer** → **primary consumer** → **secondary consumer** → **tertiary consumer**.

secondary meristem	Region of active cell division that arises from other tissues e.g. cork cambium.
secondary sexual characteristics	The physical characteristics that appear during puberty and adolescence. In the male they include the broadening of the shoulders, growth and enlargement of penis, deepening of the voice, body and facial hair, etc. In the female they include enlargement and growth of the breasts, growth of body hair under arms and pubic regions.
secretion	The release of something 'useful' from a cell or gland e.g. hormones, enzymes.
seed	A mature/ripened ovule consisting of an embryo and food store which can give rise to a new individual. See cotyledon for diagram.
selective breeding	The selection of individual plants or animals that have desirable traits and crossing/mating them in the hope that the progeny produced will have the combined, desired characteristics.
selective weedkiller	A chemical which kills certain types of weeds only e.g. any weedkiller containing 2,4-D will kill only broad leaved plants (e.g. daisy, dandelion, clover, plantain, etc.) in lawns but will not harm the grass.
selectively permeable	See semi-permeable.
self-fertilise	(Usually plants only.) The union of haploid male and female gametes produced by the same gland. Compare cross-fertilise.
self pollination	The transfer of pollen from the anther of the stamen of one flower to the stigma of the carpel of the same flower or another flower on the same plant.
semen	Seminal fluid produced by the male that contains the sperm, together with fluid from the seminal vesicles, prostate gland and Cowper's gland.

semi-circular canal(s)	Structure found in the inner ear that detects movement of the head and controls the balance of the body. See the ear for diagram.
semi-lunar valve(s)	Found in the heart only at the entrance to both the aorta and pulmonary artery. They prevent the backflow (reflux) of blood into the heart. See the heart for diagram.
seminal vesicle	Manufactures seminal fluid (semen). This fluid nourishes the sperm and allows them swim. See the male reproductive system for diagram.
semi-permeable	Also selectively permeable or partially permeable. Refers to the cell membrane. Allows certain molecules or ions to pass through but prevents others. Note: the cell wall is fully permeable i.e. it allows all molecules or ions to pass through.
sense organs	Organs, containing receptors that detect changes in the internal and external environments, e.g. ear (hearing), eye (sight), nose (smell), skin (touch) and tongue (taste). See the skin for diagram.
sensitivity	Having a keen awareness of or affected by stimuli from the external and internal environments.
sensory neuron	Afferent neuron. Pick up and carry messages from sense organs (receptors) to the central nervous system (CNS). Cell body at end of a short branch to one side of the axon, outside CNS. See neuron for diagram.
sensory system	The system of sense organs in the body.
sepal	Part of a flower, usually green in colour. Protects the flower when it is a bud. See flower for diagram.
septum	A wall that divides a structure in two parts e.g. nasal septum in nose, septum in heart, etc. See heart for diagram.

serum	This is the blood plasma without fibrinogen. Serum does not clot.
sex chromosome	One of a pair of chromosomes which contains the genes that helps determine the sex of an individual e.g. in humans XX = female, XY = male. The X chromosome is larger than the Y chromosome.
sex hormones	Hormones produced in the gonads which promote the development and maintenance of the secondary sexual characteristics and structures. They also prepare the female for pregnancy, and aid in gamete formation. Males produce testosterone and females produce oestrogen and progesterone.
sex-linkage	Genes carried on the sex chromosomes (i.e. X and Y chromosomes) are sex linked. They are transmitted together so the phenotype is related to the sex of the individual. Those carried on the part of the X chromosome that has no corresponding part on the Y chromosome (i.e. non-homologous part of X chromosome) are X-linked. Examples of such genes are those controlling haemophilia and red/green colour-blindness in humans.
sexual intercourse	Occurs in three phases: 1. Arousal: in the male blood spaces in penis become filled with blood, results in penis enlarging and becoming erect. In the female arousal causes the erection of the clitoris and secretion of mucus by the wall of the vagina. 2. Copulation: the erect penis enters the vagina. The movement of the penis inside the vagina causes semen to be ejaculated from the penis into the top of the vagina. Semen contains sperm and glandular secretions. 3. Orgasm: the pleasurable experience felt during ejaculation.

sexual reproduction	**Reproduction** involving the production, transfer and union of sex cells or gametes and development of the embryo. Two parents involved (plant or animal), one male and one female.
shoot	That part of a plant that grows above the ground. Young branch or sucker, new growth of plant. See shoot tip.
shoot tip	End of shoot containing meristematic tissue. Area or zone of active cell division and growth.
sieve plate	Perforated structure at end of sieve tube cell.
sieve tube cell	Elongated cells in phloem tissue, placed end to end and separated by a perforated cell plate so that the cytoplasm of one cell is continuous with the cytoplasm of the adjoining cell. See phloem for diagram.
silage effluent	Stream flowing from a silo containing compressed grass. High in nitrates, should not be allowed to enter rivers. See also eutrophication.
single-celled protein	Micro-organisms used as human or animal food.
sinoatrial node	(S-A node.) See pacemaker.
skeletal muscle	Attached to bones. Responsible for movement. Striated = transverse stripes running across muscle. Voluntary muscle.
skeleton	Rigid framework of bones, internal or external, to support, move and protect the body. See human skeleton for diagram.
skin	

skull	Composed of (a) the cranium, which protects brain and eyes, and gives shape to the head and (b) the jaws that contain the teeth used in feeding.
slightly movable joints	These are joints found between the vertebrae and also between the bones of the hips. See vertebral column for diagram.
small intestine	Ileum. That part of the alimentary canal between the duodenum and the caecum. Most digestion and absorption occurs here. Inner walls not smooth, thrown into folds. Folds covered by thousands of tiny projections called villi. Villi increase surface area for release of enzymes and absorption and transport of digested food.

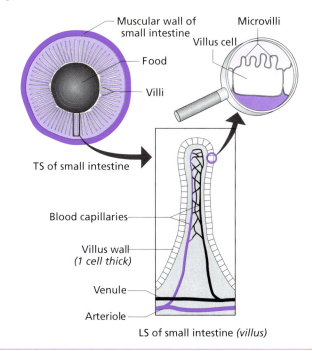

Muscular wall of small intestine
Microvilli
Villus cell
Food
Villi
TS of small intestine
Blood capillaries
Villus wall
(1 cell thick)
Venule
Arteriole
LS of small intestine (villus)

smallpox	Contagious disease caused by a virus. Symptoms include fever and pimples that may leave permanent scars. Eradicated almost worldwide today as a result of the work of Edward Jenner.

smog	A change in the atmosphere caused by pollutants produced by fuel combustion mixing with the air. An urban problem due to the high number of vehicles. A mixture of smoke and fog.
smooth muscle	Found in arteries, veins and intestines. Usually contract slowly and rhythmically e.g. peristalsis. Involuntary muscle.
soil fertility	The ability of the soil to produce abundant, fruitful crops due to the presence of all the nutrients necessary for growth.
solute	Substance that is dissolved in a liquid.
solution	A mixture of a solute and a solvent. Solute is the substance that will be dissolved and solvent is the liquid that will do the dissolving, usually water.
solvent	A liquid in which the solute is dissolved.
somatic cell	Of or pertaining or relating to any cell that is not a reproductive cell i.e. not a cell producing gametes or spores.
specialisation (zone)	Another name for the zone of differentiation. See root cap for diagram.
species	Group of animals or plants that can interbreed and produce viable, fertile offspring. Members of a species share the same characteristics and differ only in minor details. Each member of a species is unique.
specific defence system	Composed of the organs of the immune system that react to the presence of antigens by producing antibodies. This confers immunity which usually lasts a long time.

sperm	A mature male haploid gamete, also called a spermatozoan. Human sperm can survive for about 48 hours inside the female reproductive system.

Head —
Acrosome (contains digestive enzymes)
Nucleus (contains 23 chromosomes)

Middle piece —
Collar (containing many mitochondria)

Tail —
Flagellum (causes sperm to swim)

sperm duct	Also vas deferens. Tube through which sperm travel from their place of storage (the epididymis) to the urethra in humans. See male reproductive system for diagram.
sphincter muscle	A ring of muscle surrounding a tube or opening that closes when the muscle is contracted e.g. pyloric sphincter in stomach, anal sphincter in rectum, etc. Controls the movement of substances through it. See liver for diagram.
spina bifida	Congenital deformity of the spinal column associated with a deficiency of folic acid during pregnancy.
spinal cord	Is a nerve going from the brain down the inside of the spinal column. It acts as a reflex centre and also conducts messages to and from the brain.
spindle	During metaphase of mitosis and meiosis it is the thread-like structure formed from pole to pole in the cell for chromosomes to become attached to. See metaphase for diagram.

spiral	1. Classification grouping of bacteria. 2. Like a coiling circular stairway; helical; like the thread on a screw.
spleen	Large organ found in most mammals on the left hand side of the stomach. It has many functions: produces lymphocytes and antibodies, eliminates worn out erythrocytes, acts as a blood storage organ.
spongy bone	This is the inner layer of bone found at the ends of long bones. Gives strength and rigidity. Is less dense than compact bone. May contain red marrow that produces blood cells and yellow marrow that contains fat-storage tissue. See bone for diagram.
sporangiophore	A stalk-like structure (hypha) of a fungus, which grows perpendicular to the substrate and bears the sporangia or spores. Functions in reproduction. See *Rhizopus* for diagram.
sporangium	(Plural = sporangia.) A structure in which spores are produced e.g. in *Rhizopus*. See *Rhizopus* for diagram.
spore	A reproductive cell or group of cells that can give rise to a new individual without the production of gametes. Has no food store.
stamen	Structure in a flower that produces pollen, consists of anther and filament. See flower for diagram.
starch	A polysaccharide made of repeated glucose molecules. It is the main method of storing food in plants. Produced in photosynthesis in stroma of chloroplasts.
stationary phase	The third phase in a growth curve. Birth rate = death rate due to competition for food, space and the build-up of toxic wastes.
stem	That part of a plant that usually grows upwards from the soil. Possesses leaves and buds. Its main functions are support, conduction (xylem and phloem) and sometimes storage. See leaf for diagram.

stem cell	Simple (unspecialised) cell from which specialised cells will develop.
stem tuber	Swollen underground stem tip. Buds (eyes) produce new shoots e.g. potato.
sterile	Free from all types of micro-organisms. Or Incapable of producing fertile gametes.
sterilise	To make something free of micro-organisms. Or To remove the ability to produce gametes e.g. vasectomy or tubal ligation in humans.
sterility	The state of being sterile. See sterile.
sternum	The breastbone. See human skeleton for diagram.
steroid	Any one of a number of compounds that have a common molecular structure. Steroids have important physiological actions in the body and can be manufactured artificially to treat disease. Naturally occurring steroids include the sex hormones, bile salts and progesterone.
stigma	That part of the carpel on which the pollen grains must land if pollination is to be successful. See flower for diagram.
stimulus	(Plural = stimuli.) Any change in the environment (internal or external) that causes a cell or organism to respond.
stock	The stump, butt or main trunk onto which the scion is grafted.
stolon	Fungal hyphae which grow horizontally on surface of substrate (e.g. bread). See *Rhizopus* for diagram.

stoma	(Plural = stomata.) A tiny opening in the upper or lower epidermis of a leaf through which gases pass (gaseous exchange). Surrounded by two guard cells.

Stoma open   Stoma closed

stomach	A sac-like structure found in the abdomen at the end of the oesophagus and before the duodenum. Muscular walls help churn food and the process of digestion is helped by the production of gastric enzymes and hydrochloric acid. See the digestive system for diagram.
stroma	The material in a chloroplast lying between the grana. See chloroplast for diagram.
style	The 'neck' of a carpel on top of which is the stigma and below is the ovary. Upon germination of the pollen grain the pollen tube grows down through the style. See flower for diagram.
substrate	The substance on which an enzyme acts to produce the product. See enzyme-substrate complex for diagram. Or Nutrient medium for bacteria. Or Surface for attachment and/or nutrition e.g. bread for *Rhizopus*. See *Rhizopus* for diagram.
sugaring	A method of food preservation. In high concentrations it removes water from bacterial cells by osmosis (plasmolysis). Kills all bacteria.

suppressor T-cells	Lymphocytes that can suppress the immune response of B-cells and other T-cells. Prevent the immune system from over-reacting i.e. they shut off the antibody production when an infection is under control. Regulate the immune system.
surface tension	Force caused by the pull of the surface film of a liquid, because of the cohesive forces (cohesion) between the molecules, which results in minimising the surface area of the liquid. See growth curve for diagram.
survival phase	The final phase in a growth curve. Growth settles at a level that the environment can support or, in the case of bacteria, endospores may be formed due to the adverse conditions. See growth curve for diagram.
suspensory ligament	Attaches the ciliary body to the lens of the eye. It helps to control and adjust the shape of lens. See the eye for diagram.
suture	Junction of two bones of the skull by fusion of their edges.
sweat	Perspiration, or liquid produced by sweat glands; contains water (95%), salt (2%), carbon dioxide (3%), urea (very small %) and lactic acid (very small %). See perspiration.
sweat gland	Exocrine gland, excretory organ – produces sweat. See the skin for diagram.
symbiosis	Mutualism: two organisms of different species living together or one within the other to their mutual advantage, e.g.: (a) alga and fungus in a lichen (b) nitrogen-fixing bacteria in the nodules of leguminous plants (legumes): the bacteria make nitrogen compounds needed by the plant and the plant makes carbohydrates and other food material needed by the bacteria (c) bacteria living in the colon produce vitamin B_2 and vitamin K. The body absorbs these vitamins.

synapse

Neurons do not touch each other. They are separated by small gaps = synapses. Allow one-way conduction only of an impulse. When an impulse reaches the terminal button or synaptic knob, it causes acetylcholine to be released. This crosses the gap and sets up an impulse in the next neuron. Enzymes then break down acetylcholine and the gap is cleared for the next transmission.

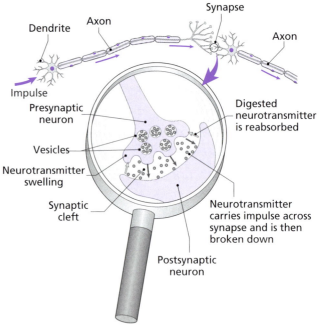

Synapse

Dendrite

Axon

Axon

Impulse

Presynaptic neuron

Digested neurotransmitter is reabsorbed

Vesicles

Neurotransmitter swelling

Synaptic cleft

Neurotransmitter carries impulse across synapse and is then broken down

Postsynaptic neuron

synaptic cleft

The gap between two neurons.
See synapse for diagram.

synaptic knob

(Neurotransmitter swelling). A swelling at the end of an axon.

synovial fluid

Liquid produced by the synovial membrane that helps to lubricate a joint and reduce friction.
See synovial joint for diagram.

synovial joint	Where bones meet and are held together by ligaments, with cartilage and synovial fluid between them. Types of synovial joints include hinge joints and ball and socket joints.

Ligament Tendon Muscle

Synovial membrane

Cartilage

Patella

Synovial fluid

Ligament

synovial membrane	Bag-like structure (sac) surrounding a freely movable joint (synovial joint) e.g. elbow, knee. The membrane to bones at either side of the joint: produces synovial fluid. See synovial joint for diagram.
synthesis(e)	Combining separate small units to form a more complex structure e.g. elements to molecules, amino acids to proteins, monosaccharides to polysaccharides, etc.
system	Group of organs concerned with one function e.g. digestive system. See pulmonary circulation for diagram.
systemic circulation	The pathway of the blood from the heart to the body and back to the heart again. Compare pulmonary circulation.
systole	Contraction phase of cardiac cycle during which the chambers discharge the blood. Compare diastole.

Tt

T-cells	See T-lymphocytes. See also helper T-cells; killer T-cells; suppressor T-cells.
T-lymphocytes	Do not produce antibodies but act in one of four processes as helper T-cells, killer T-cells, suppressor T-cells or memory T-cells.
tap root	Swollen root for food storage. Not a reproductive organ. Biennial plants e.g. carrot, turnip. See root for diagram.
tapetum	In the anther of the stamen, a layer of nutritive cells surrounding the pollen sacs. See pollen mother cell for diagram.
target organ	A structure or organ in a body on which a hormone has a specific effect.
tarsals	The bones of the ankle joint. See leg for diagram.
taxonomy	The science of naming and classifying organisms. See classification.
telophase	The final stage of mitosis and meiosis when the new nuclei are formed and the cytoplasm of the cell divides, by the formation of a cleavage furrow in animal cells and the formation of a cell plate in plant cells. This s followed by interphase.

Nuclear membanes form

Chromosomes elongate to form chromatin

tendon	A non-elastic connective tissue that joins muscle to bone. See synovial joint for diagram.

tension	Being stretched or the pressure exerted by a gas expanding.
tertiary consumer	A carnivore which obtains its nutrition from animals, usually the fourth number of a food chain, e.g. primary produce → primary consumer → secondary consumer → teritary consumer.
test cross	Used to determine the genotype of an individual. The organism with the unknown genotype is crossed with an individual of known genotype – ALWAYS the double recessive or homozygous recessive. Noting the results of the F_1 phenotype will allow you to determine the F_1 genotype. This shows the type(s) of gametes produced by the unknown individual from which you can determine whether the unknown is homozygous or heterozygous.
testa	Seed coat. Outer covering of seed formed from integuments.
testis	(Plural = testes.) The male organ that at maturity produces gametes (sperm). Also an endocrine gland that produces the hormone testosterone. See male reproductive system for diagram.
testosterone	A hormone produced by the testes. Stimulates male secondary sexual characteristics.
thalamus	Part of the brain under the cerebrum responsible for pleasure and pain. See brain for diagram.
theory	Idea based on principles or assumptions; reasonable explanation.
theory of evolution	States that all living organisms have descended from common ancestors, and over millions of years simple organisms have developed into more complex ones.
theory of natural selection	Formulated by Charles Darwin. All species have, by a series of gradual changes, evolved from pre-existing different species.

thermo-receptors	Sensory receptors that respond to changes in temperature. See receptor.
thigmotropism	The growth response of a plant to touch e.g. tendrils.
thoracic	Of or pertaining or relating to the thorax.
thorax	That part of the body between the neck and abdomen containing the lungs, heart and ribs.
thrombocytes	Also platelets. Small fragments of larger cells formed in marrow. No nuclei. Important in blood clotting.
thymine	One of the nitrogenous bases found in DNA – a pyrimidine.
thymus	Endocrine gland found in the lower neck and upper chest. Produces hormones and lymphocytes prior to puberty. It degenerates after sexual maturity. See endocrine gland for diagram.
thyroid	Large endocrine gland situated below larynx around trachea. Has two lobes. Absorbs iodine and joins it to a protein to form thyroglobulin and stored. Converted to thyroxine when needed. See endocrine gland for diagram.
thyroid stimulating hormone (TSH)	Hormone produced by the pituitary gland which stimulates the thyroid gland to release thyroxine. Thyroxine increases cell metabolism.
thyroxine	Hormone produced by the thyroid gland. Stimulates cells to use oxygen $\Rightarrow$ controls rate of respiration i.e. controls metabolism. Excess thyroxine causes Grave's disease. A deficiency in an adult causes myxoedema whereas a deficiency in a child causes cretinism.
tibia	Shin bone of leg from knee to ankle. See leg for diagram.

tissue	Group of cells with a similar function e.g. • plant tissues: meristematic tissue and vascular tissue • animal tissues – muscular tissue and nervous tissue.
tissue culture	A method for growing individual cells in a container of sterile nutrient medium to which hormones and growth substances may have been added. This process is used in cancer research and plant propagation. See also micro-propagation.
tissue fluid	See extracellular fluid.
toxin	Noxious biochemical substance (poison of animal or vegetable origin) formed in the body by a pathogenic organism.
trace element	Elements required in very small amounts by the body but essential for normal growth and development e.g. • for plants: boron, copper, zinc • for animals: most minerals and vitamins.
trachea	Tube through which air enters a body, animal, insect etc. See the respiratory system for diagram.
tracheids	See xylem tracheids.
trait	(Genetics) character or distinguishing feature of an organism e.g. eye colour, height, sex, etc.
transcription	The making of RNA from a single strand of DNA.
transfer RNA	See tRNA
transgenic organism	An organism whose genetic content has been changed by adding a gene by genetic engineering.
translation	The making of a protein from a mRNA by translating the genetic information it contains to a sequence of amino acids.

translocation	In plants, it is the movement of substances (food) from one place/location (e.g. leaf) to another (e.g. root tip or shoot tip). Or In genetics, it is a type of chromosome mutation involving the exchange of parts of chromosomes between non-homologous chromosomes.
translucent	Allowing light to pass through but cannot see through e.g. similar to frosted glass.
transpiration	The loss of water vapour from the surface of a plant. See transpiration stream.
transpiration stream	The flow of water through a plant, from root hair cells to stomatal openings in leaves, as a result of the loss of water by transpiration. See Dixon-Joly theory.
transverse section	See TS.
triceps	Large muscle at the back of the upper arm. See antagonistic muscle pairs for diagram.
tricuspid valve	A valve in the heart composed of three flaps, preventing the backflow of blood, between the right atrium and right ventricle. See the heart for diagram.
triglyceride	Smallest lipid consisting of three fatty acid molecules and one glycerol molecule.

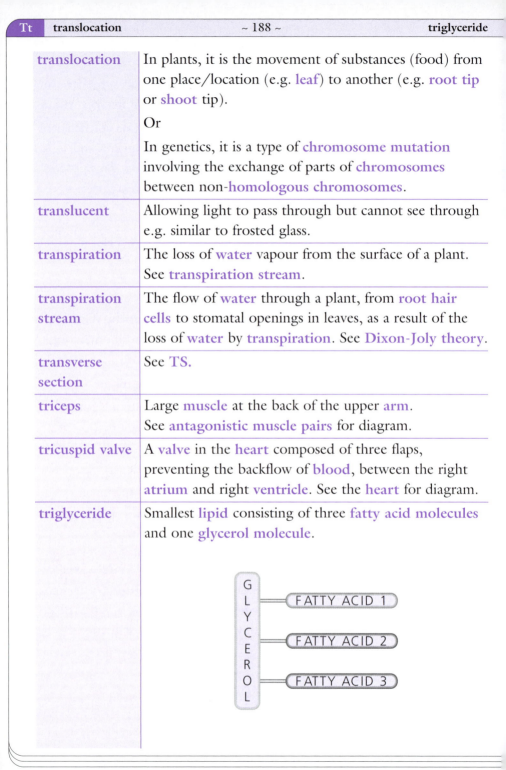

triplet	A sequence of three nitrogenous bases of mRNA that codes for a specific amino acid.

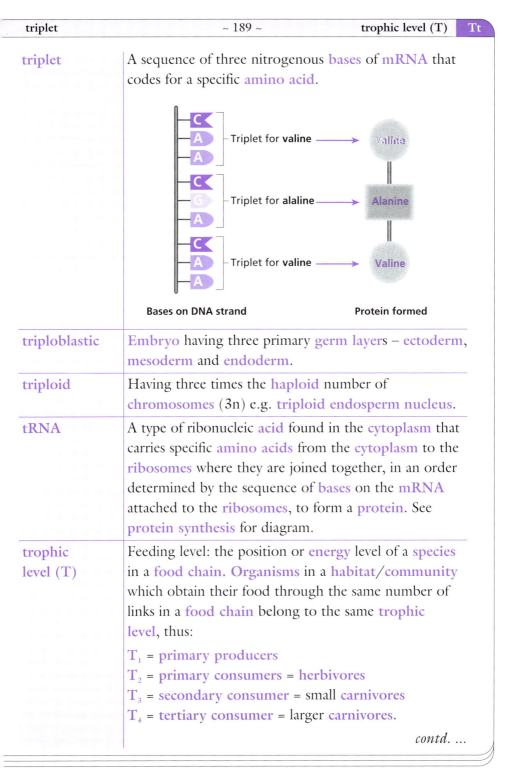

Bases on DNA strand   **Protein formed**

triploblastic	Embryo having three primary germ layers – ectoderm, mesoderm and endoderm.
triploid	Having three times the haploid number of chromosomes ($3n$) e.g. triploid endosperm nucleus.
tRNA	A type of ribonucleic acid found in the cytoplasm that carries specific amino acids from the cytoplasm to the ribosomes where they are joined together, in an order determined by the sequence of bases on the mRNA attached to the ribosomes, to form a protein. See protein synthesis for diagram.
trophic level (T)	Feeding level: the position or energy level of a species in a food chain. Organisms in a habitat/community which obtain their food through the same number of links in a food chain belong to the same trophic level, thus: T_1 = primary producers T_2 = primary consumers = herbivores T_3 = secondary consumer = small carnivores T_4 = tertiary consumer = larger carnivores.

contd. ...

Sample food chains and their trophic levels				
Trophic level	**1st**	**2nd**	**3rd**	**4th**
Stage	Producer	Primary consumer (herbivore)	Secondary consumer (carnivore)	Tertiary consumer
Hedgerow examples	hawthorn	caterpillar	robin	hawk
Seashore examples	plankton	barnacle	whelk	crab

trophoblast	Surface layer of cells of blastocyst, part of which forms the placenta. See blastocyst for diagram.
trophoblastic villi	See villus.
tropic response	The growth of part of a plant as a result of an external stimulus (e.g. light, gravity, etc.) promoting the production of an auxin. See tropism.
tropism	The growth response of part of a plant to an external unidirectional stimulus: • positive tropism: plant grows towards stimulus • negative tropism: plant grows away from stimulus.
true-breeding	When self-fertilisation gives rise to the same traits (phenotypes) in all the progeny for successive generations i.e. the genotype of the parents is homozygous.
TS	Transverse Section. View of a section of an organism cut at right angles to the long axis e.g. cross section of stem or root showing vascular structures.
TSH	See thyroid stimulating hormone.
tubal ligation	Cutting and tying, or removal of part of, the fallopian tubes for sterilisation of the female. Compare vasectomy.

tube nucleus	One of the nuclei in the pollen grain in seed plants. The tube nucleus grows down through the stigma, style and into the ovule, followed by the two male gamete nuclei to enter the embryo sac through the micropyle.
tuber	A swollen end of an underground stem = stem tuber: has eyes (buds) e.g. potato. Or Swollen end of root = root tuber: has no buds (e.g. dahlia). Plants use tubers for reproducing (propagating) vegetatively.

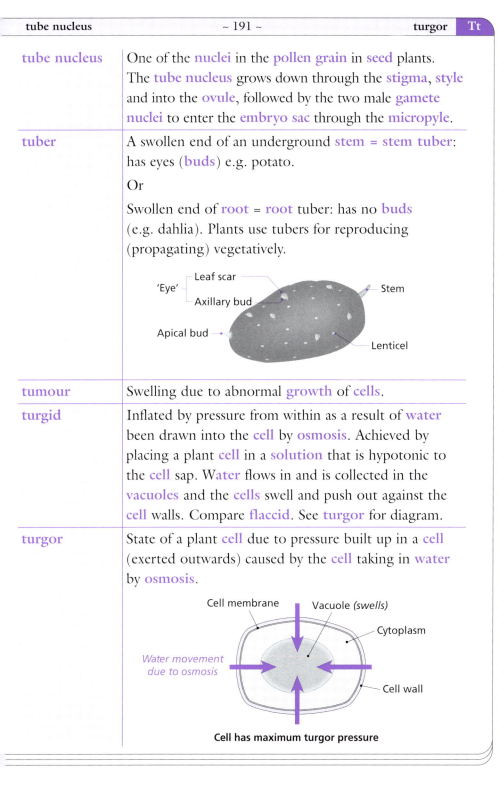

tumour	Swelling due to abnormal growth of cells.
turgid	Inflated by pressure from within as a result of water been drawn into the cell by osmosis. Achieved by placing a plant cell in a solution that is hypotonic to the cell sap. Water flows in and is collected in the vacuoles and the cells swell and push out against the cell walls. Compare flaccid. See turgor for diagram.
turgor	State of a plant cell due to pressure built up in a cell (exerted outwards) caused by the cell taking in water by osmosis.

Cell has maximum turgor pressure

Uu

ulna	**Bone** of forearm of human or other tetrapod, from wrist to elbow. On same side of **arm** as little finger. See the **arm** for diagram.
ultrafiltration	See **pressure filtration**.
ultrasound scanning	Technique used where very high frequency sound waves are passed into the body and the echo which returns is converted into a visible picture of the internal **organs**. Used on women during **pregnancy** to monitor **embryo** and foetal development.
ultrastructure	The detailed structure of something e.g. **cell**, **tissue**, or **organ**, that can only be seen using an **electron microscope**.
umbilical cord	Tube-like structure containing **blood vessels** which connects a **foetus** to the **placenta** in the **uterus**.
unicellular	Composed of one **cell**.
unsaturated fat	A **triglyceride** that has double **covalent bond**s between some of the carbon atoms. Compare **saturated fat**.
uracil	One of the nitrogenous bases found in **RNA** only in place of **thymine**. A **pyrimidine**.
urea	$CO(NH_2)_2$. A nitrogen-containing organic **compound** excreted by most **mammals**. Formed in the **liver** from **ammonia**, as a result of the breakdown (**deamination**) of excess **amino acids**, and excreted by the **kidneys**.
ureter	Duct that brings **urine** from the **kidney** to the **bladder**. See the **urinary system** for diagram.
urethra	Duct which delivers **urine** from the **bladder** to the outside; also, in the male it deposits **sperm** in the **vagina**. See the **urinary system** for diagram.

urinary system	Group of organs concerned with the production and excretion of urine i.e. kidneys, ureters, bladder and urethra.

Aorta

Vena cava

Renal artery

Right kidney

Renal vein

Left kidney

Ureter

Bladder

Uretha

Sphincter muscle

urination	Process of emptying the bladder of urine.
urine	Fluid produced by the kidneys, stored in the bladder (urinary bladder) and discharged through the urethra. Fluid contains waste products of metabolism e.g. water, urea, salts, hormones, etc.
uterus	The womb. Holds developing embryo during pregnancy. See female reproductive system for diagram.

Vv

vaccination	Process of injecting a vaccine (which acts as an antigen) into a person in order to induce immunity (produces antibodies against the antigen).
vaccine	A suspension of living, dead or attenuated pathogens which act as an antigen causing the body to produce antibodies that render the body immune to infection by the specific pathogens.
vacuole	Only found in plant cells. Fluid-filled cavity in the cytoplasm containing aqueous solutions/salts. Maintains turgidity of cell by osmosis, helps in osmoregulation. See cell for diagram.
vagina	Holds the erect penis during sexual intercourse. Also birth canal i.e. new baby arrives into the world from the uterus through here. See female reproductive system for diagram.
valve(s)	A flap of tissue which permits the flow of a substance in one direction only e.g. as in heart and veins. Valves prevent the backflow of blood. See the heart for diagram.

to the heart

Vein

Skeletal muscle *(relaxed)*

Valve

Direction of blood flow

Vein is compressed
(blood pressure increases)

Muscle (contracts)

Valve prevents backflow

variation	Differences between members of a species, group or population. Only those variations (changes) that can be inherited or passed on are advantageous as they will accumulate over millions of years and give rise to new species. Variations are the 'functional units' of evolution.
vas deferens	See sperm duct.
vascular	Pertaining or relating to the circulatory system in animals. Or Woody conducting elements (xylem and phloem) in plants.
vascular bundle	A longitudinal strand of conducting tissue made of xylem and phloem.
vascular system	A transport system. In animals it is the blood/circulatory system and in plants it is the water (xylem) and food (phloem) conducting system.
vascular tissue	Refers to the tissues involved in the transport of substances i.e. xylem, phloem in plants and blood in animals.
vasectomy	Cutting and tying, or removal of part of, the vas deferens for sterilisation of the male. Compare tubal ligation.
vasoconstriction	Method of thermal insulation employed by humans to conserve heat in cold conditions. The blood vessels nearest the surface are constricted (made narrower) so less blood flows to the body surface and less heat is lost to the atmosphere.
vector	An organism that transmits a pathogen i.e. a carrier of a disease, infection or gene from one organism to another e.g. the tsetse fly is the carrier of malaria.

vegetative propagation	**Reproduction** not involving **seed** e.g. **rhizomes**, **corms**, **tubers**, **bulbs**, cuttings, layering, **grafting**, **budding** and **tissue culturing**. See **asexual reproduction**.
vein	Thin-walled **blood** vessel that carries **blood** to the **heart** slowly at low pressure. Has a large **lumen**. Also a three layered wall: • outer layer: non-elastic **fibres**. • middle layer: elastic **fibres** and **muscles** (thin layer) • inner layer: **endothelium**, one **cell** thick. Has **valves**, which prevent **blood** flowing backwards. **Blood** flows steadily – no pulse. Collagen Involuntary muscle and elastic fibres Endothelium Lumen **TS of artery** **TS of vein**
vein (leaf)	A strand of **vascular tissue**, usually visible, in the **leaf** of a plant.
vena cava	A large **blood vessel** (**vein**) that brings **blood** from the body back to the right **atrium** of the **heart**. See **heart** for diagram.
ventral	Of or pertaining or relating to the front or lower surface.
ventral root	Pairs of projections arise from the **spinal cord**. These are called spinal roots of which there are two types: • the **dorsal root**, which has a swelling, carries **sensory neurons** into the **spinal cord** • the **ventral root** which carries **motor neurons** away from the **spinal cord**. See **reflex arc** for diagram.
ventricle(s)	A cavity or chamber of the **heart** that receives **blood** from the **auricle(s)** and passes it into the **arteries**. See **heart** for diagram.

venule	A small vein.
vertebra(e)	One of the bones that make up the spinal or vertebral column.

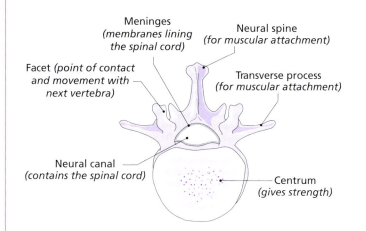

Meninges
(membranes lining the spinal cord)

Neural spine
(for muscular attachment)

Facet *(point of contact and movement with next vertebra)*

Transverse process
(for muscular attachment)

Neural canal
(contains the spinal cord)

Centrum
(gives strength)

vertebral column	Composed of 33 vertebrae; cervical (7) – neck; thoracic (12) – ribs attached; lumbar (5) – small of back; sacral (5) – hips; caudal (4) – tail.

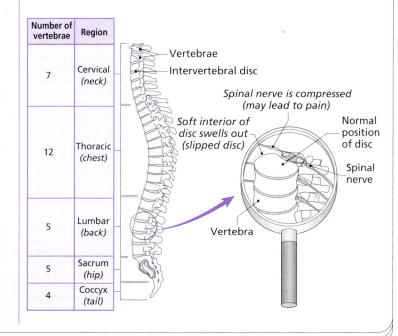

Number of vertebrae	Region
7	Cervical *(neck)*
12	Thoracic *(chest)*
5	Lumbar *(back)*
5	Sacrum *(hip)*
4	Coccyx *(tail)*

Vertebrae

Intervertebral disc

Spinal nerve is compressed *(may lead to pain)*

Soft interior of disc swells out *(slipped disc)*

Normal position of disc

Spinal nerve

Vertebra

vertebrate	Animal having a spinal or vertebral column, member of the phylum *vertebrata*, e.g. humans, birds, reptiles, fish, etc.
Vertical Section	See VS.
vestigial structure	A tissue or organ whose size, due to loss of function, has been reduced over evolutionary time.
villus	(Plural = villi.) Small finger-like extensions/processes found in the small intestine which increase the surface area for secretion and absorption. Also trophoblastic (chorionic) villi, these help to form the placenta.
virus	(Plural = viruses.) Non-cellular micro-organisms, made up of a protein coat and one type of nucleic acid (DNA or RNA). Obligate parasites = can only multiply inside a living cell. Three types of virus: spheres, rods and bacteriophages. Cause disease e.g. foot and mouth, rabies, polio, influenza, common cold, AIDS, etc.
vitamin	An essential organic catalyst of the metabolic processes. Needed in small amounts, cannot be produced in the body. Must be supplied continuously and in sufficient quantities. Some are water soluble e.g. vitamin B and vitamin C and some are fat soluble e.g. vitamin A, vitamin D, vitamin E and vitamin K.

vitamin C	Ascorbic acid, water-soluble. Involved in tissue repair and growth of connective tissue especially skin and blood vessels. Deficiency causes scurvy. Source: citrus fruits, green vegetables, tomatoes.
vitamin D	Calciferol, fat soluble, essential for the absorption of calcium that is needed for bone and teeth formation and their healthy maintenance. Stored in the liver. Deficiency causes rickets in children and osteomalacia in adults. Source: liver, eggs, sunlight.
vitreous humour	Jelly-like substance found between lens and retina of the eye. Maintains eyeball shape. See the eye for diagram.
voluntary	Under conscious control.
VS	Vertical Section. View of a section of an organism cut in the direction of the long axis (see LS) or cut perpendicular to the horizontal plane.

Ww

warfarin	Chemical used in large quantities in rat poison. Prevents the clotting of blood and rats die from excessive bleeding (haemorrhaging). Used in small quantities in human medicines to prevent blood clotting.
water	Essential for life. • Makes up 70–95% of cell mass. Component of cytoplasm and body fluids. • An excellent solvent, most chemical reactions take place in water. • One of the reactants in photosynthesis. • A product of respiration. • Component of sweat and urine. • Involved in osmosis, helps control the shape of cells. • Needed for germination. • Has a high specific heat capacity = a lot of energy required to change its temperature.
weathering	Partial disintegration or discoloration of rocks and stones as a result of exposure to the air.
white blood cell(s)	Have a nucleus but no definite shape. Protect the body against disease. There are two main types, lymphocytes and monocytes, each with different functions.

Vacuole
(containing bacteria or viruses)

Membrane

Pseudopodia

Nucleus

white matter	Found in the spinal cord. Consists of the axons of the neurons. See reflex arc for diagram.
whorl	Circle or ring of leaves arising from a stem, or rings of flower parts within the flower.
wilting	Drooping of the leaves of a plant resulting from loss of turgidity due to lack of water.

Xx

X chromosome	One of the **sex chromosomes**. See **sex chromosome** for diagram.
xylem	**Water** and **mineral** conducting **tissue** in plants. Forms the woody **tissue** and provides mechanical support. See **xylem tracheids** and **xylem vessel** for diagram.
xylem tracheids	Long dead **cells**, containing **lignin** for support. Tapered at both ends, pits in the walls to allow sideways movement from **cell** to **cell** of **water** and **minerals**. See **xylem vessel**.

Tapered end —

Cell wall —

Spiral lignin —

Pits —

xylem vessel	Tubes of dead **cells**, no end walls. Form continuous tube, wider than **xylem tracheids**. Have **lignin** in spiral bands for thickening. Transport **water** and **minerals** upwards from **root** to **stem**. See **xylem tracheids**.

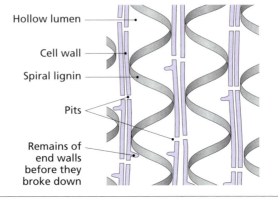

Hollow lumen —

Cell wall —

Spiral lignin —

Pits —

Remains of end walls before they broke down

Yy

Y chromosome	One of the sex chromosomes. See sex chromosome for diagram.
yeast	Unicellular fungus, ovoid or spherical cells found singly or in groups (e.g. on fruit skins such as apple and grape). Thin cell wall, granular cytoplasm, nucleus and storage vacuoles.

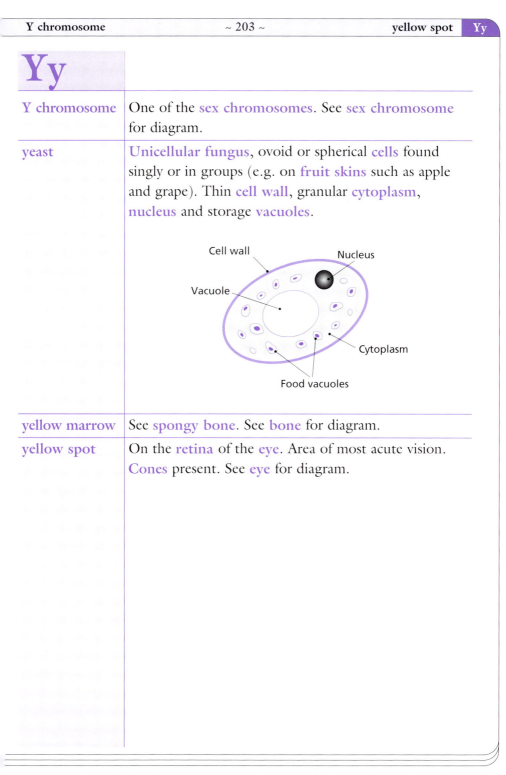

Cell wall
Nucleus
Vacuole
Cytoplasm
Food vacuoles

yellow marrow	See spongy bone. See bone for diagram.
yellow spot	On the retina of the eye. Area of most acute vision. Cones present. See eye for diagram.

Zz

zooplankton	Animal plankton.
zygospore	A thick-walled resting spore with a resistant coating to survive adverse conditions, produced from the union of two gametes.
zygote	A diploid cell resulting from the union of two haploid gametes – a fertilised egg.

18500948